Principles and Methods in Receptor Binding

NATO ASI Series

Advanced Science Institutes Series

A series presenting the results of activities sponsored by the NATO Science Committee, which aims at the dissemination of advanced scientific and technological knowledge, with a view to strengthening links between scientific communities.

The series is published by an international board of publishers in conjunction with the NATO Scientific Affairs Division

A	Life Sciences	Plenum Publishing Corporation
B	Physics	New York and London
C	Mathematical and Physical Sciences	D. Reidel Publishing Company Dordrecht, Boston, and Lancaster
D	Behavioral and Social Sciences	Martinus Nijhoff Publishers
E	Engineering and Materials Sciences	The Hague, Boston, and Lancaster
F	Computer and Systems Sciences	Springer-Verlag
G	Ecological Sciences	Berlin, Heidelberg, New York, and Tokyo

Recent Volumes in this Series

Series A: Life Sciences

Principles and Methods in Receptor Binding

Edited by

F. Cattabeni

Institute of Pharmacology and Pharacognosy
University of Urbino
Urbino, Italy

and

S. Nicosia

Institute of Pharmacology and Pharmacognosy
University of Milan
Milan, Italy

Plenum Press
New York and London
Published in cooperation with NATO Scientific Affairs Division

Proceedings of a NATO Advanced Study Institute on
Principles and Methods in Receptor Binding,
held September 8–18, 1982,
at Urbino, Italy

Library of Congress Cataloging in Publication Data

NATO Advanced Study Institute on Principles and Methods in Receptor Binding
    (1982: Urbino, Italy)
    Principles and methods in receptor binding.

    (NATO ASI series. Series A, Life Sciences; v. 72)
    "Proceedings of a NATO Advanced Study Institute on Principles and Methods
in Receptor Binding, held September 8–18, 1982, at Urbino, Italy"—T.p. verso
    Includes bibliographic references and index.
    1. Radioligand assay—Congresses. 2. Cell receptors—Congresses. 3. Drug re-
ceptors—Congresses.4. Binding sites (Biochemistry)—Congresses. I. Cattabeni,
Flaminio. II. Nicosia, Simonetta. III. North Atlantic Treaty Organization. IV. Title.
[DNLM: 1. Binding sites—Congresses. 2. Ligands—Congresses. QU 34 N277p
1982]
QP519.9.R34N38   1982   574.87'5   83-27248
ISBN 0-306-41613-1

©1984 Plenum Press, New York
A Division of Plenum Publishing Corporation
233 Spring Street, New York, N.Y. 10013

Printed in the United States of America

PREFACE

One of the major advances in the understanding of the mechanism of action of hormones, neurotransmitters and drugs had arisen from the hypothesis that the physiological or pharmacological responses are triggered by their interaction with specific cell components, termed receptors. However, the presence of receptors has been inferred from indirect data, and only recently has it been possible to study the kinetics of the interaction between drug and receptors directly, through the so called "binding technique."

This NATO-ASI on "Principles and Methods in Receptor Binding" was devoted mainly to the following aspects of the study of receptors: the principles underlying the use of the binding technique; the mathematical models necessary to interpret binding data; the application of binding methods to specific receptors; and, finally, a few selected examples of coupling between receptors and physiological responses.

In the chapters of this book, special interest is devoted to the analysis of the simplest models for the interaction of receptors with their ligands (either hormones or neurotransmitters or drugs). The graphical techniques used to analyze the data from binding experiments are extensively discussed, together with the statistics that have to be used in binding analysis. Moreover, the basic concepts to analyze binding data using a personal computer are presented.

The factors which should be considered when setting up a binding assay (such as choice of ligand and incubation conditions, preparation of tissue, termination of incubation) are discussed. Receptors which are described in some detail are those for GABA and benzodiazepines, ß-adrenergic ligands, dopamine and acetylcholine.

Recent trends in the study of receptors are mentioned: the stimulating hypothesis of co-transmitters as regulators of ligand-receptor interaction is discussed, and the recent technique of incorporation of receptors into artificial membranes is described. Finally, the coupling of receptors to ion channels and to adenylate cyclase is treated.

The Institute has been made possible by the efforts of many dedicated individuals. We are expecially grateful to the Organizing Secretary, Dr. I. Ceserani; to the members of the Organizing Committee, Dr. E. Coen, Dr. G. Lombardelli, Dr. D. Oliva and Dr. G. Peruzzi; to Dr. M.P. Abbracchio, Dr. F. Benfenati, Dr. N. Brunello, Dr. M. Cimino, Dr. M. Lombroso and Dr. D. Oliva, who have made the practical demonstrations successful.

In addition to the generous support and sponsorship of NATO, the Institute was co-sponsored by the Italian Society of Pharmacology and the University of Urbino.

Financial support was provided by the Municipality of Urbino and the following companies: Beckman Analitica, Ciba Geigy, Fidia, Glaxo, I.S.F., Nen Italiana, Sandoz and Zambeletti.

Finally, we wish to greatly thank the collaboration of the personnel of the "Opera Universitaria" who made the stay at the College pleasant for all the participants.

                                        F. Cattabeni
                                        S. Nicosia

CONTENTS

LIGAND BINDING DATA ANALYSIS:

THEORETICAL AND PRACTICAL ASPECTS

Peter J. Munson

Laboratory of Theoretical and Physical Biology
National Institute of Child Health and Human Development
NIH, Bethesda, MD

INTRODUCTION

Popularity of the ligand-receptor binding study as a means of
investigating the mechanism of action of drugs, hormones, and neurotransmitters has grown considerably during the last decade. This
growth is largely due to the availability of radiolabeled compounds,
the simplicity of the method and its versatility of application.
The low "energy barrier" of the methodology and the relative ease
with which experiments may be performed may sometimes result in
comparably small effort being spent on the data analysis. The
most commonly used approach for data analysis is the preparation
of a Scatchard plot of the B/F ratio vs. Bound ligand concentration.
In the simplest case, the data will lie along a straight line from
which one may abstract the binding affinity from the slope and the
binding capacity from the [Bound] axis intercept. However, the data
may sometimes show a curvilinear relationship which complicates the
analysis considerably. Even when the Scatchard plot is linear, there
are numerous pitfalls in the interpretation of the analysis. Resulting misunderstanding and even abuse of the Scatchard analysis is
wide-spread in the literature, to the extent that a prominent
physical chemist was recently prompted to write that "in most cases
... the conclusions drawn from the Scatchard graph are completely
untenable"[1]. Although this statement may be overly pessimistic,
the sentiment is well founded. Ligand binding studies have been and
will continue to be an important source of information about the
mechanism of receptor action. Yet investigators are well advised
to be cautious in the interpretation of their analysis. In this
paper we review some of the more common errors in the analysis,
and discuss how routine use of a well-designed computerized, model
fitting program may aid in avoiding them.

PITFALLS IN SCATCHARD PLOT ANALYSIS

The first problem arising with the Scatchard plot analysis is the subjectivity of drawing a straight line through a scatter of data points in the Scatchard plot. Since no single straight line will fit all of the points perfectly, the investigator may be tempted to choose the line which best supports his favorite hypothesis. Of course, when the data have little scatter, the overall distortion due to subjectivity will be small. On the other hand, when the scatter is large or there is an occasional "outlier" point, significant problems arise. With a properly trained intuition (a rare and precious thing) the investigator can minimize these problems, but it is quite difficult to prove the correctness of one's intuition.

In order to eliminate subjectivity from the analysis, some investigators have attempted to apply linear regression to the Scatchard plot. Unfortunately, this simple approach does not completely solve the problem. First there are not one but two linear regressions: first B/F vs [Bound], and regression in the transposed plot, [Bound] vs B/F. The results of these two regressions will not be identical and one is again left with the problem of choosing between them. Several assumptions must be validated before the use of linear regression may be considered appropriate. In particular, the independent or "X" variable must be known precisely, without error. In the Scatchard plot, neither Bound nor B/F may serve as the independent variable since both are measured quantities. In practice, application of linear regression to the Scatchard plot is less satisfactory than simply fitting the data manually with a straight edge. The use of linear regression is therefore not recommended, except for getting rough initial estimates of the parameters.

The next problem with the analysis of the Scatchard plot arises with the application of an inappropriate or incorrect model to the data. One frequently finds examples where the points of the Scatchard plot are fit with a straight line, although the points display a definite curvilinear relationship. This approach has sometimes been justified on the basis that it is standard practice for investigations in a particular system. However, such evidence of lack of fit should not be ignored. At best, fitting a curved Scatchard plot with a straight line gives a biased estimate of the binding affinity and capacity. At worst, such estimates are meaningless.

Substantial curvature may be due to the presence of two distinct classes of binding sites. If so, one can attempt to recover the binding parameters for each class, a process referred to as "resolving" a Scatchard plot into its components. One may resolve the curve graphically, but the process is so tedious that it is

rarely completed correctly. An <u>incorrect</u> simplification of
the method treats the upper and <u>lower</u> "linear" segments of the
curve as identical with the high and low affinity components.
Unfortunately, use of this method is widespread and leads to
substantial errors in the calculation of values. The true components are represented by graphical asymptotes to the hyperbola.
The slopes and intercepts of these asymptotes are proportional to
the binding affinities and capacities for each component. Alternatively, one may arrive at correct estimates of these components
using the slopes of the upper and lower regions of the curve itself
with the use of the "limiting slopes" technique[2]. In most situations however, it is more expeditious to apply a computerized
curve fitting program directly to the data.

Finally, investigators are cautioned not to ascribe too much
significance to parameters such as B_{max} which require extrapolation
of the curve outside the data range. Although on a Scatchard
plot the [Bound] axis intercept appears to be only a short extrapolation outside the range of the data, it is in fact extrapolation
to an infinite free ligand concentration. This becomes more
apparent when the data are plotted on a [Bound] vs log [Total] or
log [Free] scale. Most investigators are already aware of the
dangers inherent in extensive extrapolation. For instance, the
binding model itself may no longer apply at very high free ligand
concentration because of second or third class of binding sites.

COMPUTERIZED ANALYSIS OF LIGAND BINDING DATA

Because of the significant difficulties encountered in the
analysis, investigators frequently turn to computer programs for
the analysis of binding data. One should not do this too casually,
however, as there are many incorrect programs in existence and even
the best programs may occasionally give "garbage" output. Therefore
the investigator must be able to critically evaluate the program
and its output, and determine that an appropriate analysis technique
has been applied.

A proper statistical analysis begins with the definition of the
dependent and independent variables, i.e. the coordinate system for
curve-fitting. The independent variable should be known without
significant measurement error. With ligand binding experiments one
may usually take total ligand concentration, the concentration of
ligand initially present in each tube or vial. This value is known
quite precisely. Free concentration by contrast is usually calculated by subtraction of [Bound] from [Total] and therefore may
contain substantial measurement error. Bound ligand concentration
at equilibrium becomes the response or dependent variable in the
analysis. Use of the [Bound] vs [Total] coordinate system is
preferable to (B/F vs [Bound]) or ([Bound] vs log [Free]) or

(1/[Bound] vs 1/[Free]) or any other transformation of the raw data
because it isolates the measurement error into a single variable and
the statistical nature of this error is easily characterized.

Because measurement error is not constant over the observation
range, but tends to increase with the response value, statistical
weighting is required in the nonlinear regression. Many computer
programs facilitate the use of weighting by including a mathematical formula which predicts the expected variation of an observation based its magnitude. Thus, larger measured values of [Bound]
are predicted to have larger standard errors.

The binding model, i.e. the mathematical representation of
the chemical binding reactions, must be selected with some care.
Since one must often consider a variety of binding models, the
computer program ought to formulate and fit each candidate model in
turn and finally select the one most consistent with the data.
Models are frequently chosen from the following list: one binding
site model with parameters K and R; two-site model with parameters K_1, R_1, K_2 and and R_2; various models for negative
or positive cooperativity; models for the interaction of
several ligands with either one or more classes of receptors
(general "m by n" model); non-competitive or un-competitive
inhibition as well as competitive binding of two ligands and a
single receptor.

With each of these models, one should include a parameter N
for non-specific binding, to account for apparent non-saturable
classes of binding sites present on albumin, binding globulins or
other proteins. The importance of fitting this parameter has been
demonstrated [3,4]. The simpler approach of subtracting nonspecific
binding from the data before fitting is less efficient and may
result in unnecessary loss of information.

After the correct model has been defined and fitted to the
data in an appropriate coordinate system with the use of weighting,
the next step is to critically and objectively analyze how well
the model fits. Again, a computer program can and should provide
a number of useful statistical tools. The first is the residual
mean square (RMS) error which gives an indication of the size of
the deviation of a typical data point from the fitted curve. In
well controlled experiments, the typical point deviates from the
curve by about 5 or 10%. Statistics for the number of runs above
and below the curve, and the serial correlation of the residuals
are useful in characterizing the goodness-of-fit. Most important
is the visual evaluation of the fitted curve graphically super-
imposed on the original data points. Again, this may be prepared
and presented by the computer program. For this purpose, almost
any coordinate system is useful, provided one is familiar with
its properties.

The above steps constitute a minimally acceptable computerized analysis of ligand binding data. However, a number of other
features which are also desireable. First is the ability to pool
results from more than one binding assay. Results are often
collected on a number of different assays and an average value of
affinity and capacity may be desired. One may fit each assay
separately and simply average the results. However, fitting all
the data simultaneously will generally give superior results.
Simultaneous fitting of data allows one the opportunity to test
if the affinities changed between experiments. If not, one may
find the binding capacities for each experiment, assuming the same
binding affinity. In some cases both affinity and capacity will
change between experiments. One may test for this by comparing
the goodness-of-fit statistics for the simultaneous data fit with
and without constraints on the binding parameters. The F-test
may be used to determine whether any of these parameters have
changed significantly and to assign the appropriate p-values.

The output of a well-designed computer program should include
not only the "answer" i.e. the value of parameters of interest but
also the standard error or precision of the estimate frequently
expressed as %CV (percent coefficient of variation). Typically,
binding affinities and capacities for a single site model may be
determined to within 10 or 20%. One should also obtain the correlation between estimates of pairs of parameters. For instance,
between the parameter for affinity (K) and for capacity (R) there
is often a strong negative (-0.95) correlation. Thus, spuriously
high values for K may be associated with unusually low values for
R and vice versa. An appreciation for the correlations between K
and R is helpful when comparing results of two experiments. Finally,
a flexible computer program should allow for more sophisticated
experimental designs, especially for cross-displacement studies
involving 2 or more ligands binding to 1, 2, or more classes or
receptor sites. This capability has been used to advantage in the
study of pituitary receptors, subtypes of the alpha and beta
-adrenergic receptors[5] and more recently, classification of kappa
mu and delta opiate receptors[6].

PRACTICAL ASPECTS OF PREPARING THE DATA FOR ANALYSIS

From a statistical point of view it would be ideal if one
could fit the raw data, e.g. counts of radioactivity directly.
However, it is often more convenient to first convert the data
into standardized units. For instance, specific activity may be
used to convert bound counts to mass of Bound ligand. While such
conversions at first seem straightforward, the details must be
explicitly specified, since even small perturbations introduced
at this step may have a substantial influence on later affinity
estimates. In addition to the bound counts from the gamma counter,
certain ancillary information should be collected:

For each tube one must know the initial concentration of unlabeled
ligand. One must know the specific activity of any labeled
ligands present. Unfortunately the nominal specific activity
provided by the manufacturer is only approximate and may contribute
to substantial errors in binding parameters. A more reliable measure
of specific activity is obtained by performing a "self-displacement"
study where the ED_{50} of tracer displaced by unlabeled ligand is compared to the ED_{50} of tracer displaced by itself, i.e. a second displacement curve is formed by adding only increasing amounts of label.
The ratio of the two ED_{50}s is the specific activity. One should
adjust the value of total labeled ligand concentration for the proportion of tracer which is "unbindable", that is, for the biological
purity of the label. The process of radio-labeling often damages
the ligand rendering it incapable of binding to receptor. This
is especially true for proteins. Thus, some of the measured counts
may be attached to unbindable, "dead" ligand. Alternative measures
of ligand purity such as HPLC should be utilized to assess the
bindability of the tracer. Efficiency of the gamma or beta
counter must be determined in order to convert observed radioactive
counts (cpm) into actual disintigrations (dpm) occuring in the tube.
This value is typically around 50%, but may vary significantly
between machines or experiments especially if a scintillation
cocktail is used.

Background counts and nonspecific binding should be carefully distinguished. Background counts are the number of counts
which are detected by particular counter machine in the absence of
radio-active tracer, i.e. cosmic radiation, residual contamination
of the counter, etc. Non-specific or non-saturable binding is
measured by adding the tracer ligand to the tissue preparation in the
presence of a large excess cold concentration, the effect of which
is to saturate all "specific" binding sites. Bound radioactivity
under these conditions is attributed to nonsaturable binding, e.g.
to such things as albumin or serum binding proteins or even to
glass test tube walls.

If the level of background is known precisely, it may be subtracted from the original data provided it represents a only small
proportion of the total counts. Non-specific counts, on the other
hand, are better treated during the curve-fitting step as a parameter
rather than pre-subtracting them from the raw data. This approach
will improve the overall efficiency of the data analysis, and result
in more accurate standard error estimates for the other fitted
parameters.

Ideally, bound and free ligand will be separated perfectly before
measurement of bound counts. In reality, small amount of free ligand
may contaminate the bound fraction and/or only a portion of the bound
ligand may actually be recovered as the bound fraction, especially
using filtration where the filter may only collect less than 50% of

the cells. If these proportions can be estimated, a further
correction of the data is possible.

A few auxiliary parameters must also be determined. If
weighting is to be used in the non-linear curve fitting process,
weights must be assigned to each individual data point, often
using a weighting or variance model. The model will predict, on
the basis of previous experience, the expected variation in any
given observation. It is often empirically appropriate to assume
that the variance of an observation is proportional to the square
of its value, that is there is a constant relative error in measurements. Here, the variance model becomes: $Var (Y) = a_2 Y^2$. A more
detailed description of the variance of an observation might include
a constant term which dominates at low response levels: $Var(Y) =
a_0 + a_2 Y^2$. An empirical description of a variance model may be
obtained by methods described elsewhere[7]. Choice of variance
model parameters may sometimes have a significant influence on
the statistical tests based on the curve fit.

Finally one must account for the particular design of the
experiment, which has an effect on the type of data analysis to
be performed. Two commonly used designs may be termed the "hot
plus cold" case and the "hot only" case. In the first case, a
constant amount of labeled ligand is added to each tube with
varying, increasing amounts of unlabeled ligand added to the
series, yielding a "displacement curve". In the second case,
increasing amounts of hot, labeled ligand either in the presence
or absence of a large excess cold concentration are present in
each tube, giving a "saturation curve". Fundamentally, both
designs are the same since the receptor "sees" a changing concentration of ligand, and it doesn't distinguish labeled from
unlabeled moleculed ligand. From this point of view, both experiments may be described in terms of the total ligand concentration and bound ligand concentration in each tube, and can be
analyzed in the same way.

INFLUENCE OF EXPERIMENTAL ARTIFACTS

Systematic errors which enter during the pre-processing step
may be the dominant source of error in estimates of binding affinity
and capacity. Therefore, the true uncertainty in the parameter
values is probably several times larger than the standard errors
printed out by most computer programs. Such standard errors (or
%CV's) rely only on the average scatter of the points around the
curve, and cannot take into account such things as the unknown
uncertainty in the specific activity. Typically specific activity,
tracer purity, counting efficiency, counter background and efficiency
of separation are taken as fixed known quantities, yet systematic
or random errors in these quantitities will dramatically influence
the results. To examine the effect of such experimental artifacts,

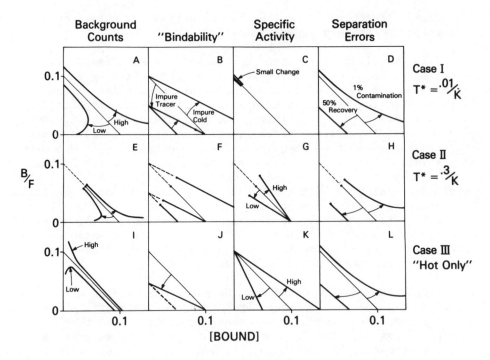

FIG. 1. Effect on Scatchard plot of presence of six undetected
artifacts. The "true" curve is identified in each panel with a
B/F intercept of 0.1 and a Bound axis intercept of 0.1 The true
curve is distorted as shown in the presence of each artifact,
under conditions of infinitesimal tracer (T^*=0.01/K, first row),
substantial tracer (T^*=o.3/K, second row), or using labeled ligand
only ("Hot Only", third row). The magnitude of the artifactual
factors are as follows: Counter background is present as 0.01%
(Panel A), 0.1% (Panel E) of Total counts, yielding the upper
curves in both cases. The lower curves (Panels A and E) result
from over correction by similar amounts. Counter background
represents 10% of the Total counts in the lowest dose in the "Hot
Only" case (Panel I), yielding the upper curve. Overcorrection by
a similar amount yields the lower curve in Panel I. 50% tracer
purity is assumed for the lower pair of curves in panels B and F
and the middle line in J. 50% purity of unlabeled ligand is
assumed for the upper "Impure Cold" curve shown in panels B and F.
Both tracer and cold ligand are 50% pure for the second line from
the bottom, panels B and F. The dashed line in panel J arises
from a 50% pure tracer along with 50% overestimated specific
activity. A 50% underestimated (Higher than stated true value) or
overestimated (Low) specific activity is assumed for the curves in
panels C, G, and K. Finally, recovery of only 50% of the Bound
ligand yields the lower curve in panels D, H and L, while
contamination of Bound by 1% of Free ligand gives rise to the
upper curves in the same panels (taken from ref. 8).

we have simulated the effects of variations in these quantities[8].
The effect on the Scatchard plot changes with the magnitude and
type of error made and also with the particular design chosen.
Case I and Case II refer to a "hot + cold" or displacement curve
experiment using a constant amount of labeled ligand. In Case I
the concentration of labeled ligand represents 0.01 K_d. In Case
II, the tracer concentration is 30 times higher or 0.3 K_d, resulting
in substantial self-displacement even before the binding study
is begun. Though not optimal, many investigators are forced to
use a Case II design because of low specific activity. Case III
is the "hot only" saturation experiment where increasing amounts
of labeled ligand are added to each tube in the assay.

     Figure 1 demonstrates the effect of artifacts on the shape of the
Scatchard plot. In each figure, the true, un-distorted binding
is shown as a straight line Scatchard plot with intercepts of 0.1
on the B/F axis and 0.1 on the [Bound] axis. The other lines or
curves show the effect of an undetected artifact in the data.
Table 1 summarizes the effects on the estimated parameter values.
In figure 1A, the upper curve ("High") shows the effect of not
correcting for background counts for a Case I experiment. Background
counts mimic the effect of non-specific binding or of a second
class of binding sites. If background counts were overestimate
and too large a value subtracted from the data, the lower curve
("Low", Fig. 1A) would be obtained. Which may give the appearance
of positive cooperativity. Thus, even with as little as 10 cpm
undetected background out of 100,000 total counts could shift the
apparent binding capacity by a substantial amount. In comparison,
figure 1E shows that using more tracer hence more total counts
makes the experiment much less sensitive to background count
rate. The effect of background counts in the "hot only" design
is much less apparent at the high dose region of curve, making
this design is more desireable for measuring B_{max} accurately.

     Conversely, the "hot only" design is more sensitive to inacu-
racies in the specific activity. Figs. 1C, 1G and 1K show the
effect of over- or under-estimation of specific activity used in
calculation. For the Case I, "hot plus cold" experiment (Fig. 1C)
the net effect is very small since the labeled proportion of ligand
is very small. In Fig. 1G (Case II), a larger percentage of
the ligand is labeled giving specific activity a more substantial
overall effect. If specific activity is overestimated by the
investigator (the nominal value is higher than the true value)
the apparent curve will be lower than it should be resulting in a
comparable underestimation of the binding affinity but not binding
capacity. Compare this to Fig. 1K (Case III) where a 50% over-
or under estimation of specific activity affects both affinity
and binding capacity in an inverse manner. The upper curve ("High")
results from underestimation of specific activity. If specific
activity is questionable, and binding

Table 1. Effect of experimental artifacts on apparent binding parameters taken from a Scatchard plot.

	Case I			Case II			Case III		
	"K"	"R"	"KR"	"K"	"R"	"KR"	"K"	"R"	"KR"
Undetected background	↓	↑↑	↑	↓	↑↑	~	~	~	↑↑
Overcorrected background	↑	↓↓	↓	↑	↓↓	~	~	~	↓↓
Impure tracer	--	↓	↓	--	↓	↓	↓	--	↓
Impure cold	↓	↑	--	↓	↑	--			
Impure tracer & impure cold	↓	--	↓	↓	--	↓			
Underestimated true specific activity	~	--	~	↑	--	↑	↓	↑	--
Overestimated true specific activity	~	--	~	↓	--	↓	↑	↓	--
Impure tracer and over-estimated s.a.							--	↓	↓
Contamination of Bound with Free	↓	↑↑	~	↓↓	↑↑	~	↓	↑↑	~
Low recovery of Bound	--	↓	↓	--	↓	↓	--	↓	↓

↑ indicates increase; ↓ , decrease; ~ , little change; --, no change. "K" refers to the apparent affinity, the negative of the "average" slope, "R" refers to Bound axis intercept, "KR" refers to B/F axis intercept.

capacity is to be measured accurately, one is advised not use the
Case III, "hot only" approach.

Bindability or purity of the labeled or unlabeled ligand
will also dramatically change the location of the apparent binding
curve. In Case I and Case II (Fig 1B, 1F) note that impure tracer
alone reduces the apparent intercepts for both B/F and [Bound]
axis by a factor of 2. Thus, the binding capacity is changed but
the binding affinity remains constant. If only the cold ligand
were impure, the apparent B/F intercept would be unchanged but
the [Bound] axis intercept and hence the apparent binding capacity would increase by a factor of 2. The third possibility is
that both tracer and cold are less than fully pure. The resulting
curve shows a reduced B/F axis intercept with an unchanged
[Bound] axis intercept. Binding affinity is the only parameter
affected in this situation. In Case III (Fig. 1J), if only 50%
of the tracer ligand is "bindable", a reduced binding affinity
results without change in the binding capacity. However, impure
labeled ligand is often accompanied by overestimated specific
activity especially when the specific activity has been measured
in a self displacement study. The result of this combination is
a reduced binding capacity measurement with no change in the
apparent affinity (Fig. 1J, dashed line).

Finally we consider the effect of errors in separation of
bound and free ligand. The contamination of the bound fraction
with 1% of the free ligand gives an effect similar to non-specific
binding (Figs. 1D, 1H, and 1L). Conversely, if the method of
separation is inefficient so that only a portion of the bound
ligand is recovered in the "bound fraction", further errors will
result. Reduced recovery efficiency results in equivalent reduction
in the binding capacity value with very little effect on the
binding affinity.

The effects on the shape of the plot will be exaggerated
with initial binding (B/F) substantially higher than 0.1. If
this value is in the range of 0.5 to 1.0 artifacts that raise or
lower the initial binding (e.g. inefficient recovery) may also
result in a perceptibly curved plot (concave down or up) even
though only a single class of sites is present.

The important influence of specific activity, bindability
background count, etc. often goes unstated. The reported values
of binding affinity and binding capacity for many if not all
receptor systems are dependent on these ancillary parameters.
Therefore, careful investigations should focus more attention on
their correct determination and use in data analysis. Further,
when comparing values obtained in different experiments or in
different laboratories one should allow for the possibilities of
errors from these sources. In some cases, by combining

several different experimental designs, it may be possible to
detect and correct for the presence of one or more of these artifacts. Finally, these results allow one to avoid experimental
designs which are overly sensitive to particular artifacts.

References

1. I. M. Klotz, Number of Receptor Sites from Scatchard Graphs:
    Facts and Fantasies, Science 217:1247 (1982).
2. A. K. Thakur, M. L. Jaffe and D. Rodbard, Graphical Analysis
    of Ligand-Binding Systems: Evaluation by Monte Carlo
    Studies. Anal. Biochem. 107:279 (1980).
3. P. J. Munson, LIGAND: A Computerized Analysis of Ligand
    Binding Data, in: "Methods in Enzymology, Immuno-chemical
    Techniques, Part E" 92, J. J. Lagone and H. Van Vanukis,
    eds., Academic Press, New York(1983).
4. P. J. Munson and D. Rodbard, LIGAND: A Versatile Computerized
    Approach for Characterization of Ligand-Binding Systems.
    Anal. Biochem. 107:220 (1980).
5. A. DeLean, et al., Validation and Statistical Analysis of a
    Computer Modeling Method. Mol. Pharmacol. 21:5 (1982).
6. A. Pfeiffer and A. Herz, Discrimination of Three Opiate
    Receptor Binding Sites With Use of a Computerized Curve-
    fitting Technique. Mol. Pharmacol. 21:226 (1982).
7. D. Rodbard, et al., Statistical Characterization of the Random
    Errors in the Radioimmunoassay Dose-Response Variable.
    Clin. Chem. 22:350 (1976).
8. P. J. Munson, Experimental Artifacts and the Analysis of
    Ligand Binding Data: Results of a Computer Simulation,
    J. Receptor Res. 3:xxx (1983), in press.

RADIOLIGAND BINDING ASSAYS

S. J. Enna

Departments of Pharmacology and of Neurobiology
and Anatomy
University of Texas Medical School at Houston
P. O. Box 20708
Houston, Texas 77025

INTRODUCTION

Early work aimed at examining the properties of chemical transmission focused on presynaptic events such as neurotransmitter synthesis, degradation, accumulation and release. In contrast, because of technical limitations, less information was obtained in regard to the biochemical events associated with neurotransmitter receptors. This situation has changed dramatically in the recent years due to the development of techniques which make possible the identification and characterization of these sites. One of the more powerful methods for this purpose has been the radioligand binding assay (Yamamura et al, 1978). With this procedure it is possible to label selectively that portion of the synaptic membrane to which the neurotransmitter must attach in order to alter cellular activity. Because of its simplicity and power, the radioligand binding assay has become one of the more popular tools in neurobiological research. Using this procedure it is possible to study receptor site number, kinetic characteristics, pharmacological properties, and their relationship to other membrane components. Moreover, binding assays have been useful in examining the effects of drugs and disease on neurotransmitter receptors, and the procedure has been adapted for use as a simple and sensitive analytical method for measuring the concentration of neurotransmitters and drugs.

In order to use the receptor binding technique appropriately it is essential to understand the basic principles of this assay. Thus, the most fundamental assumption made with this procedure is

13

that the radioactive ligand is attaching to the receptor of interest. However, radioligands are known to adhere to various
tissue components, only some of which may be the receptor being
examined, making it necessary to define the binding site carefully.
Moreover, it must be appreciated that even subtle alterations in
the assay procedure can direct the radioligand to a different site,
leading to erroneous interpretations. The aim of the present
communication is to summarize briefly the basic principles,
criteria and applications of radioligand binding assays. Although
this procedure has been used to study hormone and transmitter receptors in a variety of organs and tissues, emphasis will be
placed on the study of neurotransmitter and drug receptors in the
central nervous system. Readers desiring more detailed information on this topic should consult any of a number of monographs
and reviews (Yamamura et al, 1978; Bylund, 1980; Schulster and
Levitski, 1980; Enna, 1983a).

BASIC PRINCIPLES OF LIGAND BINDING ASSAYS

   The Assay Procedure

     For the assay, small amounts of a radioligand are incubated
with tissue membranes in the presence and absence of a large quantity
of unlabeled ligand. After equilibrium the reaction is termina-
ted by seperating the membrane-bound ligand from the isotope in
solution. The membranes are then rinsed free of extraneous
radioligand and the bound radioactivity extracted and quantified.
The difference in the amount of radioligand bound in the absence
and presence of unlabeled substance is taken as a measure of the
amount of isotope attaching to the receptor. The basic principle of this assay is that the radioligand will bind, in a
saturable manner, to the receptor, whereas attachment to nonre-
ceptor components is less saturable and, therefore, the unlabeled species will selectively inhibit radioligand binding to
the receptor, but not influence the binding to other membrane
constituents.

     Several factors must be considered when developing a radio-
ligand binding assay. For instance, the radioligand must be
stable, selective for the receptor of interest, and it must be
possible to label the substance to a relative high specific
activity (> 10 Ci/mmole). Moreover, the manner in which the
tissue is prepared, as well as the incubation buffer, pH and
temperature will all affect the rate of equilibrium and binding
site specificity (Enna and Snyder, 1975; Enna, 1983a). The method
used for terminating the reaction may also be important with regard to the proper characterization of the binding site. Thus,
detection of displaceable binding does not, in itself, demonstrate that the ligand is attaching to the proper receptor. This
can only be proven by studies aimed at examining the specificity
of the binding site.

Binding Site Identification

Three basic criteria must be met in order to establish that the binding site represents a biologically relevant receptor (Table 1). The first of these is saturability. Thus, there is a very limited number of neurotransmitter receptors on most tissue membranes. Because of this it should be possible to readily saturate the number of displaceable binding sites. This is examined by measuring the binding in the presence of increasing concentrations of radioligand in the presence of a fixed concentration of unlabelled ligand. Saturation is indicated when the amount of displaceable binding becomes constant as a function of radioligand concentration (Figure 1). In contrast, the amount of radioactivity bound in the presence of unlabeled ligand normally increases in a linear fashion since the number of "nonspecific" binding sites is virtually infinite. This latter value, termed the blank, represents attachment to constituents other than the receptor binding site. The point at which the displaceable binding plateaus represents the saturable binding component. In the example shown, the concentration of binding sites (saturation point) is approximately 28 pmoles of radioligand per mg of protein.

The second criteria for demonstrating binding site specificity relates to the distribution of the site inasmuch as displaceable binding should only be found in those tissues known to contain biologically active receptors. If physiological tests have revealed that the receptor is found only in certain areas of the central nervous system, then displaceable binding should not be detectable in other brain areas, nor should it be found in peripheral tissues. Likewise, if the receptor is known to be a component of the synapse, then displaceable binding should be most enriched in that subcellular fraction containing the highest concentration of synaptic membranes (Burt, 1978).

Table 1. Basic Criteria for Radioligand
          Binding Assays

---

1. Saturability

2. Distribution

3. Substrate Specificity

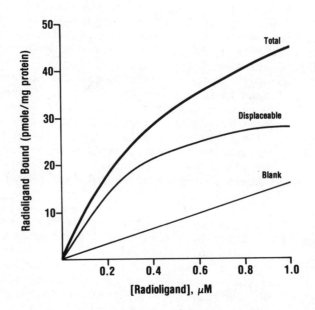

Fig. 1.   Representative curves for a radioligand binding site
          saturation study. Radioligand binding was measured as
          a function of increasing concentrations of isotope in the
          presence of a fixed concentration of unlabeled ligand.
          Total binding represents the amount of isotope adhering
          to the tissue in the absence of unlabeled ligand.   The
          blank represents the amount of binding in the presence
          of a fixed (saturating) concentration of unlabeled
          ligand.   Displaceable binding is the difference between
          the total binding and blank, and represents the amount
          of isotope displaced from the membrane binding sites by
          the unlabeled ligand.

The third, and perhaps the most important, criteria is substrate specificity. If displaceable binding represents attachment to the biologically relevant receptor, then only those substances known to interact with that receptor should demonstrate appreciable potency as inhibitors of radioligand attachment. For example, a recent study was undertaken to examine the binding characteristics of trazodone, an antidepressant (Kendall et al, 1983). The radioligand used to examine this question was

Table 2. Substrate Specificity of ^{3}H-THT
Binding in Rat Tissue

Substrate	IC_{50} (nM) Cerebral Cortex	Liver
Trazodone	15 ± 2	40 ± 3
Tetrahydrotrazodone	50 ± 4	90 ± 9
Serotonin	$10,000 \pm 1,650$	$100,000 \pm 4,000$
Norepinephrine	$100,000 \pm 12,500$	$1,200 \pm 76$
Metergoline	28 ± 6	$12,000 \pm 1,400$
Cinanserin	68 ± 8	——
Cyproheptadine	99 ± 12	——
Phentolamine	$2,000 \pm 350$	$4,500 \pm 500$
Prazosine	$1,500 \pm 88$	——
WB-4101	$1,300 \pm 331$	$2,800 \pm 140$

Adapted from Kendall et al (1983).

tetrahydrotrazodone (THT), an analog with a biological profile
similar to trazodone. Initial experiments revealed a displaceable
binding component for ^{3}H-THT in both cerebral cortex and liver.
Since antidepressant activity is most likely mediated in brain,
the finding of a saturable component in liver suggested that the
binding may have little relevance with regard to the mechanism of
action of this drug. However, binding site specificity studies
revealed that the membrane component labelled in brain differed
from that found in liver (Table 2). Of the substances examined,
the most potent inhibitor of ^{3}H-THT binding in either tissue was
trazodone, which inhibited 50% of the displaceable ^{3}H-THT (IC_{50})
at a concentration of 15 nM in cerebral cortex and 40 nM in liver
(Table 2). The IC_{50} for unlabelled THT was 50 nM and 90 nM in
the two different tissues respectively. Since trazodone and THT
are structurally related, the finding that both could inhibit the
binding provides little information in regard to the pharmacological specificity of the site. That is, chemical substances
of similar structure will often interact with biological membranes
at a site unrelated to the specific receptor. These sites are
more properly referred to as "acceptors", rather than receptors,
since they are simply membrane components which happen to bind a
particular chemical class. In order to demonstrate that the
labelled site has biological relevance it is essential to show
that structurally unrelated agents will interact with this component as well. Earlier studies had revealed that trazodone had
significant affinity for both serotonin and noradrenergic receptors in brain (Enna and Kendall, 1981). For ^{3}H-THT binding it
was found that structurally unrelated serotonin receptor antagonists such as metergoline, cinanserin and cyproheptadine were
relatively potent as inhibitors of ^{3}H-THT binding to cerebral
cortical membranes, with all three agents having IC_{50} values less
than 100 nM. It was also found that serotonin was the most potent
neurotransmitter examined having an IC_{50} value of 10 μM, which is 10
times greater than norepinephrine. Further evidence that cerebral
cortical ^{3}H-THT binding represents attachment to a serotonin rather
than a noradrenergic receptor was provided by the finding that
adrenergic agents such as phentolamine, prazosine and WB-4101 were
10-100 times less potent than the serotonin receptor antagonists.

Displaceable ^{3}H-THT binding had a completely different substrate specificity in liver membranes (Table 2). In this tissue
norepinephrine was some 80 times more potent than serotonin as
an inhibitor of binding. Moreover, metergoline was several
hundred-fold less active in inhibiting ^{3}H-THT binding to liver
membranes than it was in cerebral cortical tissue. These
differences indicated that ^{3}H-THT labels pharmacologically different sites in cerebral cortex and liver, with the site in the brain
appearing to be a serotonin receptor (Kendall et al, 1983). Thus,
data such as these are necessary to establish the identity of a
particular binding site.

Table 3.   Fundamental Considerations in the
Selection of a Radioligand

---

1.   Selectivity

2.   High Affinity

3.   Stability

4.   Specific Activity

LIGAND SELECTION

There are four basic considerations that go into the selection of a radioligand (Table 3). The first is that the substance
must be selective for the receptor of interest, since it is not
unusual for a drug or a neurotransmitter to have a significant
attraction for a variety of receptors. Antipsychotics such as
haloperidol are known to interact with brain dopamine receptors,
cholinergic muscarinic receptors, α-adrenergic receptors, and
serotonin receptors (Creese et al, 1977; Miller and Hiley, 1974;
Peroutka et al, 1977; Bennet and Snyder, 1975). Because of this,
neuroleptics may be used to label any one of a variety of neurotransmitter binding sites. Likewise, antidepressants are known
to attach to a number of brain receptors as well as neurotransmitter transport sites (Peroutka and Snyder, 1981). While,
ideally it is best to use a ligand known to interact with only a
single site, such specificity is exceptional. Accordingly,
efforts must be directed towards finding a substance that is
relatively selective for the receptor under investigation.

High affinity for the receptor is another prerequisite for a
radioligand. Since the number of brain receptors is extremely
small compared to the number of nonspecific binding sites, low
concentrations of radioligand must be used in order to detect the
biologically relevant component. If the affinity of the ligand
for the receptor is too low, then the concentration that must
be used will also result in the labelling of a significant number
of nonspecific sites, making it difficult to study the receptor.
In general it is best if the compound selected is known to have
an affinity for the receptor in the nanomolar range. It is for
this reason that antagonists are commonly used as radioligands
since they generally have receptor affinities greater than
agonists.

Radioligands must also be chemically stable. Three types of
stability must be considered. First, the ligand must be resistant
to change under the conditions used for labelling the substance.
Tritiation is usually accomplished either by an exchange reaction
or by the reduction of a double bond, and can be used for virtually all organic molecules. Iodination, on the other hand, is
normally restricted to those molecules possessing an aromatic hydroxyl group, such as a tyrosine residue. Of these procedures, tritium exchange is the least likely to cause an alteration in
structure since tritium is merely exchanged for hydrogen atoms
already present in the molecule. Therefore, assuming that the
conditions necessary for the exchange reaction do not damage the
molecule, it is reasonable to assume that the addition of tritium
atoms will not alter biological activity. On the other hand, by
reducing the substance in the presence of tritium the basic
structure is altered, which may cause a change in biological
activity. For this reason it is useful to have a precursor of
the desired ligand that, when reduced, forms the parent substance.
In regard to iodination, monoiodinated substances usually retain
full biological activity (Bennett, 1978). However, the addition
of an iodine atom changes the structure of the compound and
therefore may lead to a dramatic change in the biological profile.
Indeed, in some cases, a compound may become di- or tri-iodinated
which invariably alters biological activity and receptor affinity.
Because of these possibilities it is prudent to test the biological activity and potency of the radioactive compound prior to use
in a binding assay.

Another consideration is stability during storage. Some
substances, such as serotonin, are notoriously labile, rapidly
decomposing under most storage conditions. Such compounds must
be used immediately after synthesis, and precautions must be
taken to monitor purity on a continuous basis.

It is also necessary to consider the stability of the radioligand under the assay conditions since it is conceivable that,
although the ligand is stable during storage, it may be unstable
at the pH or temperature of the binding assay. To examine this
it is necessary to determine the purity of the radioligand after
the binding assay to be sure that the substance bound is the same
as that added to the incubation tubes. If a significant change
in purity is noted, it will be necessary to alter the incubation
conditions.

Specific activity is also a factor that must be considered
when selecting a radioligand. For most binding assays it is
essential to have the substance labeled to a specific activity
greater than 10 Ci/mmole. Both tritium reduction or iodination
normally achieve this level of specific activity or greater,
although tritium exchange is less predictable. Because of the

extremely small quantities of radioligand used in a binding assay, a high specific activity is necessary to maintain a ratio of specific to nonspecific binding high enough for quantifying the displaceable component. It was this limitation that prevented the development of ligand binding assays before the 1970's since, up to that time, most radioactive substances were labeled with C^{14}, an isotope which yields a specific activity in only the mCi/mmole range. This necessitated the use of relatively large quantities of radioactive substance to detect a significant amount of binding and therefore more than 90% of the total radioactivity was bound to nonspecific sites. With specific activities greater than 10 Ci/mmole however, it is possible to incubate membranes with nmolar concentrations of radioligand which, in turn, yields a specific: nonspecific ratio of two or more. Indeed, in some assays, greater than 95% of the total binding represents attachment to the receptor (Wastek et al, 1976).

TISSUE PREPARATION

The procedures used for preparing tissue for most brain neurotransmitter receptor assays are quite similar. The first step is usually preparation of subcellular fractions, since binding to neurotransmitter receptor sites is normally most enriched in the synaptosomal component. The tissue is homogenized in an isotonic sucrose solution and the suspension subjected to a low speed centrifugation to separate the crude nuclear fraction (P_1) from the synaptosomes in the supernatant (S_1). The S_1 is then centrifuged at an appropriate speed to remove the synaptosomal fragments from the suspension. This P_2 fraction is homogenized in a nonisotonic buffer or water to disrupt the membranes. This is followed by further resuspensions and centrifugations to rid the tissue of endogenous ligands. Finally, the membrane fraction is suspended in buffer and small portions (0.5 to 2 mg protein) are used for assay.

In some cases it is possible to study receptor binding in tissue homogenates. Under these circumstances the brain tissue is initially homogenized in buffer or water, and then submitted to a high speed centrifugation, after which the pellet is resuspended and centrifuged several times prior to assay. Importantly, slight modifications in the tissue preparation procedure are known to dramatically alter radioligand attachment. For example, it was found that the binding of ^{3}H-GABA to synaptic GABA receptors was greatly enhanced if the tissue was frozen prior to assay (Enna and Snyder, 1975). Likewise, serotonin receptor binding requires that the tissue be preincubated at 37° for a short period of time before assay (Bennett and Snyder, 1976). Treatment with detergents, such as Triton X-100, has been found to improve binding assays under certain conditions (Enna and Snyder, 1977). In all of these cases it is believed that the treatments increase assay sensitivity by more completely removing endogenous substances that interfere with the binding site. Since modifications in tissue preparation can influence binding site specificity and number, it is important to redefine binding site specificity if the tissue preparation has been altered.

Another important consideration relates to the amount of tissue used in the binding assay. For most assays it has been found that displaceable binding is linear with tissue concentrations up to approximately 2-3 mg protein. Beyond this the relationship ceases to be linear, possibly because of substances in the tissue that influence the stability of the radioligand, or because the amount of tissue alters the efficiency of the separation procedure. Thus, for the most precise and reproducible data, it is essential to conduct the binding assay at a tissue concentration known to be on the linear portion of the tissue binding curve.

INCUBATION CONDITIONS

The mathematics used to describe the receptor binding phenomenon are based upon the assumption that the reaction has been allowed to proceed to equilibrium (Hollenberg, 1978). Thus, anything that influences the rate of this reaction will significantly alter the binding data. Such factors are temperature, pH, and the ionic composition of the buffer. In general, the higher the incubation temperature, the more rapid will be the rate of equilibrium. On the other hand, the influence of pH is less predictable since it influences the ionic state of the ligand and binding site which, in turn, may determine the affinity of the receptor for the ligand. Therefore, experiments must be conducted in order to determine the optimal temperature and pH for the binding assay.

A variety of buffers have been found useful for conducting radioligand binding assays. Some of the more common are sodium-potassium phosphate, Tris-citrate and Tris-HCl (Enna, 1983a). In some circumstances it has been found that by varying the buffer it is possible to direct the radioligand to a different binding site. Thus, ^{3}H-GABA binding appears to represent attachment to the presynaptic GABA transport site if the assay is conducted in the presence of sodium ion, whereas binding in the absence of sodium appears to be primarily to the synaptic receptor site (Enna and Snyder, 1975). It has also been shown that sodium ion can differentially affect the affinity of opiate agonists and antagonists (Pert and Snyder, 1974). Other substances found to influence the amount or selectivity of binding are calcium, magnesium and ascorbic acid. Once again, binding site specificity must be redefined if there has been an alteration in the incubation buffer. Similarly, a change in pH or temperature necessitates a re-examination of the time to equilibrium and

substrate specificity. Moreover, because the rate of equilibrium
is temperature-dependent, it is normally unwise to conduct the
incubation and terminate the reaction at different temperatures as
this will yield inconsistent kinetic data making it difficult to
compare results.

SEPARATION TECHNIQUES

Several methods can be used to terminate the binding reaction.
Equilibrium dialysis is normally used when the radioligand affinity
is quite low (> 1 mM). For this procedure the incubation is conducted in a chamber separated into two compartments by a semiper-
miable membrane. The tissue is placed on one side of the membrane
and the radioligand on the other. Following equilibration, a
fraction taken from the tissue side is compared to the radioligand
side, with the difference in radioactivity indicating the amount
of ligand bound. This assay is useful for low affinity substances
since the bound ligand is never exposed to a rinsing medium. However, the chief disadvantage of equilibrium dialysis is that the
difference between the radioligand concentration on the two sides
of the chamber is normally quite small, making it difficult to
define the characteristics of the binding site.

The most common methods for terminating binding assays are
filtration and centrifugation (Bennett, 1978). For filtration,
the incubate is poured over a glass fiber filter that is maintained under reduced pressure. The buffer containing the free
radioactivity is suctioned through the filter, whereas the membranes remain trapped. The filters are then washed several times
with a large (10-20 ml) volume of ligand-free buffer to rid the
tissue of risidual radioactivity. The filters are placed into
scintillation counting vials, and the membrane-bound radioactivity extracted and quantified. This procedure is generally used
for radioligands having affinities in the low nmolar range since,
if the affinity is less, thorough washing may remove some membrane-bound radioligand, yielding an underestimate of receptor
number.

Centrifugation has been found useful for those substances
with affinities in the high nmolar range. After centrifuging the
samples, the supernatant is decanted and the membranes washed
rapidly and superficially with fresh buffer or water. Because
the membranes have been pelleted, the washing procedure is less
thorough than with filtration. While this method is less likely
to remove receptor bound radioactivity, it normally yields a
higher blank value. Centrifugation is often employed when initially developing a ligand binding assay and if the affinity is
found to be greater than 50 nM a filtration method will be used
to reduce the blank and increase assay efficiency.

    Another consideration when selecting a seperation technique
is the amount of radioligand bound to the filtration or centrifugation apparatus. Such binding will contribute significantly to
the blank, and may therefore complicate the accurate determination of receptor-bound ligand. It has also been shown that radioactive substances may bind to inert materials in a displaceable
manner (Cuatrecasas and Hollenberg, 1975), making it possible that
a significant fraction of the specific binding is to filters
rather than tissue. Accordingly, experiments must be conducted
to determine the extent and nature of radioligand attachment to
the materials used in the incubation and separation procedures.

DATA ANALYSIS

    The ligand binding assay reveals two fundamental characteris-
tics of receptor sites, affinity (K_d) and number (B_{max}). These
values can be determined in any of three ways (Bylund, 1980). The
first is by a saturation study in which the concentration of unlabelled ligand is held constant in the presence of increasing
concentrations of radioligand (Figure 1). The second type of experiment is an inhibition assay in which the radioligand concentration is held constant at or below the estimated K_d in the presence of increasing concentrations of unlabeled drug (Figure 2).
The third type involves a direct kinetic analysis in which the
radioligand concentration is held constant with binding determined
as a function of time. The mathematics relating to saturation and
displacement data describe a rectangular hyperbola and are equivalent to the Michaelis-Menton equation used for enzyme kinetics.
From these calculations it can be found that the equilibrium dissociation constant (K_d) is equal to the concentration of radioligand which occupies 50% of the total binding sites. Based on
this consideration, the approximate K_d of the binding site described in Figure 1 is 200 nM.

    Similar information can be obtained from a displacement ex-
periment (Figure 2). In this case the concentration of radioligand is held constant and displaceable binding is measured in
the presence of various concentrations of unlabeled compound.
When drug A is used as a displacer, 80% of the total radioactivity is inhibited. Thus, 40% displacement of the total binding
represents the relative affinity of compound A for the site,
which is approximately 25 nM. Notably, compounds B and C displace only 60% of the total radioligand. The fact that some
agents may displace more radioligand than others can be taken as
evidence that the isotope is attaching to more than one binding

site. Such a finding is not unusual, and illustrates the importance of selecting the appropriate substance for use as a blank.
For example, it is possible that the radioligand labels two distinct binding sites, one of which is displaced only by compound A
and the other having appreciable affinity for only compounds
B and C. On the other hand, it is also possible that compounds
A, B, and C all interact with the same binding site but, unlike
B and C, compound A can also displace from a second site. Further
binding site specificity studies would have to be performed to
resolve this issue.

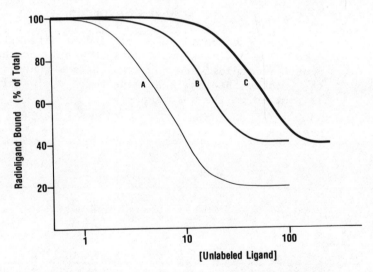

Fig. 2. Representative curves for a radioligand binding site
         displacement study. Radioligand binding was
         measured using a fixed concentration of isotope
         and increasing concentrations of unlabeled ligand.

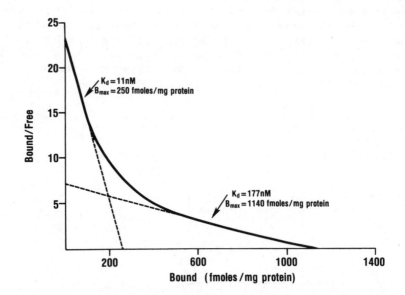

Fig. 3.   Rosenthal (Scatchard) plot of binding site saturation
          data. The ordinate represents the ratio of the
          amount of radioligand bound to the membranes to
          the unbound species. Membrane bound radioactivity
          is plotted on the abscissa.

          In order to make the most accurate determination of the K_d
and B_{max} it is necessary to analyse saturation or displacement
data using a linear transformation. Such an analysis is typically
made using a method developed by Rosenthal (1967), which was
derived from an original equation suggested by Scatchard, 1979).
In the example shown (Figure 3), a complex Rosenthal (Scatchard)
plot is illustrated. Such data suggest that there are at least
two binding sites, a high affinity component having a K_d of
approximately 11nM, and a low affinity with a K_d of approximately
177 nM. In this illustration the concentration of binding sites
for the higher affinity component is approximately one-fourth that

of the lower affinity site. As can be seen, the K_d is a function
of the slope, and the B_{max} represents the intercept on the abs-
cissa. It should be borne in mind, however, that values derived
from such data are only approximations, since the equilibrium
dissociation constant and binding site concentration for each
component is influenced by the other.

The kinetic properties of the binding site can be determined
directly by conducting an association-dissociation experiment.
Association, the rate at which the radioligand attaches to the
receptor, is determined by analysing the amount of displaceable
binding as a function of time. Dissociation, the rate at which
the ligand leaves the receptor, is determined by incubating the
samples to equilibrium, and then adding a saturating concentration
of unlabelled ligand and measuring the amount of displaced radioligand as a function of time. The ratio of the dissociation rate
to the association rate will yield a true K_d, which should be
similar to that calculated from displacement or saturation
studies.

Another type of data analysis is the Hill plot (Hill, 1910).
This reveals whether there is a linear relationship between the
amount of ligand bound and the concentration of isotope. If the
relationship is linear, this suggests that the radioligand is
attaching to a single site on the membrane. If the slope is
greater than unity, this suggests the presence of at least two
sites which bind the ligand, with attachment to one positively
influencing attachment to the other. A Hill coefficient less than
unity suggests that the multiple sites interact in a negative
manner, such that attachment to one site reduces binding to the
second component, although other interpretations are possible
(Bylund, 1980). Thus, Hill plots are useful in providing further
evidence for the presence of multiple binding sites.

APPLICATIONS

Radioligand binding assays have proven to be powerful tools
in neurobiological research. One obvious application is for drug
screening (Creese, 1978). Prior to the development of binding
assays it was necessary to examine potential receptor agonists or
antagonists using costly electrophysiological, behavioral or in
vivo biochemical techniques. Not only were these tests time con-
suming and expensive, they also required significant quantities
of test compound. However, by examining various concentrations

of test substance in a binding assay, a displacement curve can be generated (Figure 2), and the relative receptor affinity determined. In the example shown, compound B would appear to be some three-fold more active than compound C on this binding site. With this approach only very small quantities of compound are needed, and hundreds of agents can be analysed in a few hours. Of course binding assays provide no information with regard to the ability of the substance to penetrate into the brain in vivo, nor can they predict the rate or extent of metabolism.

Binding assays have also been useful in examining questions relating to receptor plasticity and development and the influence of drugs and disease states on neurotransmitter receptor sites (Enna, 1983b). For example, numerous studies have shown that chronic administration of various psychoactive drugs leads to an alteration in receptor number (Kendall et al, 1982; Sulser and Vetulani, 1977; Burt et al, 1977). These modifications may relate to some of the therapeutic and toxic effects observed following the long-term use of these agents (Kendall et al, 1982; Burt et al 1977). Other studies have shown receptor binding site alterations associated with a variety of neuropsychiatric disorders (Enna et al, 1976; Reisine et al, 1977 a & b; Bennett et al, 1979). Thus, receptor binding assays have revealed significant changes in neurotransmitter receptor number in brain tissue obtained from individuals who suffered from Parkinson's disease, Huntington's disease, or schizophrenia. Such data may provide new insights into the neurochemical abnormalities associated with these disorders.

Ligand binding assays have also been used to explore for endogenous neurotransmitters and neuromodulators (Simantov et al, 1977). For instance, when it was found that specific binding sites exist for opiates in brain tissue, it was proposed that they served as receptors for some endogenous agent. By examining the potency of various brain extracts to inhibit opiate binding it was possible to demonstrate that the brain contained a substance capable of interacting with these sites. Similar experiments are now being conducted in an attempt to discover endogenous substances that interact with the binding site for benzodiazepines, a class of anxiolytic drugs (Möhler et al, 1979).

Information relating to drug mechanisms has also been possible to obtain using receptor binding assays. In addition to determining whether drugs can directly interact with the receptor these assays can reveal indirect interactions as well. This point is illustrated by the studies showning that there is a reciprocal relationship between GABA receptor binding and the attachment of benzodiazepines to a separate, but related, site (Costa and Guidotti, 1979). These data suggest that the benzodiazepines may act at least in part, by indirectly influencing recognition site binding for GABA.

Ligand binding assays have also been adapted for use as an analytical technique to quantify neurotransmitter and drug levels in biological tissues and fluids (Enna, 1982; Ferkany and Enna, 1982). These assays provide a highly sensitive, relatively specific and rapid method for determinining the concentration of various substances. The principle of the assay is that amount of radioligand bound is a function of the unlabelled ligand present in the incubation medium. Thus, to measure the amount of dopamine receptor-active neuroleptic in blood, the ^{3}H-haloperidol binding assay is conducted in the presence of various concentrations of blood extract. By comparing the potency of the extract to inhibit binding to a standard curve for displacement of ^{3}H-haloperidol by unlabelled haloperidol, it is possible to obtain an estimate of the relative amount of neuroleptic present in the blood. A similar approach has been used to measure blood levels of a variety of drugs such as benzodiazepines, neuroleptics, and opiates (Chang and Snyder, 1978; Shibuya et al, 1977; Freedbert et al, 1979). Moreover, the method has been used to measure endogenous substances such as GABA (Enna et al, 1977).

Receptor ligand binding assays have also been important for the isolation and purification of neurotransmitter receptors (Gavish et al, 1979; Lang et al, 1979). The ability to label the receptor with a radioactive tracer makes it possible to study the receptor in a solubilized state. However, because of the paucity of any given neurotransmitters receptor in brain, and the reversibility of most ligands, only a limited amount of data has been accumulated using this technique. Nevertheless, with the development of more radioligands and new purification methods, it is likely this approach will ultimately prove successful.

SUMMARY AND CONCLUSIONS

The development of radioligand binding assays for studying neurotransmitter and drug receptors represents a significant advance in technology. The ability to study neurotransmitter receptor characteristics at the molecular level makes it possible to more accurately define the mechanism of receptor action. While a major advantage of radioligand binding assays is their technical simplicity, it is absolutely essential to have a working knowledge of the fundamental principles of these assays to properly interpret the data. The most fundamental assumption made is that the radioactive substance is attaching specifically to the receptor of interest. However, ligand binding is influenced significantly by the way the tissue is prepared, the incubation conditions, and the stability of the radioligand. When conducted properly, these assays can provide new information about drug interactions at receptor sites, as well as yielding insights into the influence of drugs and disease states on neurotransmitter receptor systems. Moreover, radioligand binding assays may be used as a primary screen in drug development, as a tool to aid in the search for new endogenous compounds of biological importance, and as a simple analytical tool for measuring the tissue concentration of neurotransmitters and drugs. Since new ligands are being developed regularly, it seems likely that radioligand binding assays will continue to provide information of fundamental importance to both the basic scientist and clinician.

ACKNOWLEDGEMENTS

    Preparation of this manuscript was made possible, in part, by the support of USPHS grant NS-00335, a Research Career Development Award, and by a grant from the National Science Foundation (BNS-8215427). The author thanks Ms. Doris Rayford for her excellent secretarial assistance.

REFERENCES

Bennett, J. P., 1978, Methods in binding studies, in "Neurotransmitter Receptor Binding", H. I. Yamamura, S. J. Enna and M. J. Kuhar, eds., Raven Press, New York.

Bennett, J.P., Enna, S.J., Bylund, D.B., Gillin, J. C., Wyatt, R. J. and Snyder, S. H., 1979, Neurotransmitter receptors in schizophrenic frontal cortex. Arch. Gen. Psychiat., 36:927.

Bennett, J. P. and Snyder, S. H., 1975, Stereospecific binding of d-lysergic acid diethylamide (LSD) to brain membranes: relationship to serotonin receptors. Brain Res., 94:523.

Bennett, J. P., and Snyder, S. H., 1976, Serotonin and lysergic acid diethylamide binding in rat brain membranes: relationship to serotonin receptors. Mol. Pharmacol., 12:373.

Burt, D. R., 1978, Criteria for receptor identification, in "Neurotransmitter Receptor Binding", H. I. Yamamura, S. J. Enna, and M. J. Kuhar, eds., Raven Press, New York.

Bylund, D. B., 1980, "Receptor Binding Techniques", Society for Neuroscience, Bethesda, Maryland.

Burt, D. R., Creese, I. and Snyder, S. H., 1977, Antischizophrenic drugs: chronic treatment elevates dopamine receptor binding in brain. Science, 196:326.

Chang, R., and Snyder, S. H., 1978, Benzodiazepine receptors: labeling in intact animals with ^{3}H-flunitrazepam. Eur. J. Pharmacol., 48:213.

Costa, E., and Guidotti, A., 1979, Molecular mechanisms in the receptor action of benzodiazepines. Ann. Rev. Pharmacol. Toxicol., 19:531.

Creese, I., 1978, Receptor binding as a primary drug screening device, in "Neurotransmitter Receptor Binding", H. I. Yamamura, S. J. Enna and M. J. Kuhar, eds., Raven Press, New York.

Creese, I., Schneider, R., and Snyder, S. H., 1977, ^{3}H-Spiroperi-
    dol labels dopamine receptors in pituitary and brain. Eur.
    J. Pharmacol., 46:377.
Cuatrecasas, P., and Hollenberg, M. D., 1975, Binding of insulin
    and other hormones to non-receptor materials: saturability,
    specificity and apparent "negative cooperativity". Biochem.
    Biophys. Res. Comm., 62:31.
Enna, S. J., 1982, Radioreceptor assays for neurotransmitters
    and drugs, in "Handbook of Psychopharmacology", Vol. 15, L.
    L. Iversen, S. D. Iversen and S. H. Snyder, eds., Plenum
    Press, New York.
Enna, S. J., 1983a, Principles of receptor binding assays: the
    GABA receptor, in "Membrane-Located Receptors for Drugs and
    Endogenous Agents", A. Reid, G. M. Cook and D. J. Morre,
    eds., Plenum Press, New York.
Enna, S. J., 1983b, Receptor regulation in, "Handbook of Neuro-
    chemistry", Vol. 6, A. Lajtha, ed., Plenum Press, New York.
Enna, S. J., Bird, E. D., Bennett, J. P., Bylund, D. B., Yamamura,
    H. I., Iversen, L. L., and Snyder, S. H., 1976, Huntington's
    chorea: changes in neurotransmitter receptors in the brain.
    N. Engl. J. Med., 294:1305.
Enna, S. J. and Kendall, D. A., 1981, Interaction of antidepres-
    sants with brain neurotransmitter receptors. J. Clin.
    Psychopharmacol., 1:12S.
Enna, S. J. and Snyder, S. H., 1975, Properties of γ-aminobutyric
    acid (GABA) receptor binding in rat brain synaptic membrane
    fractions, Brain Res., 100:81.
Enna, S. J., and Snyder, S. H., 1977, Influences of ions, enzymes,
    and detergents on GABA receptor binding in synaptic membranes
    of rat brain. Mol. Pharmacol., 13:442.
Enna, S. J., Wood, J. H., and Snyder, S. H., 1977, γ-Aminobutyric
    acid (GABA) in human cerebrospinal fluid: radioreceptor
    assay. J. Neurochem., 28:1121.
Ferkany, J. W. and Enna, S. J., 1982, Radioreceptor assays, in
    "Analysis of Biogenic Amines", G. B. Baker and R. T. Couts,
    eds., Elsevier/North Holland, New York.
Freedberg, K. A., Innis, R. B., Creese, I., and Snyder, S. H.,
    1979, Antischizophrenic drugs: differential plasma protein
    binding and therapeutic activity. Life Sci., 24:2467.
Gavish, M., Chang, R. S., and Snyder, S. H., 1979, Solubilization
    of histamine H-1, GABA and benzodiazepine receptors. Life
    Sci., 25:783.
Hill, A. W., 1910, The possible effects of the aggregation of
    the molecules of hemoglobin on its dissociation curves. J.
    Physiol., 40:iv.
Hollenberg, M. D., 1978, Receptor models and the action of neuro-
    transmitters and hormones, in "Neurotransmitter Receptor
    Binding", H. I. Yamamura, S. J. Enna and M. J. Kuhar, eds.,
    Raven Press, New York.

Kendall, D. A., Slopis, J., Duman, R., Stancel, G. M. and Enna, S. J., 1982, The influence of hormones on drug-induced modifications in neurotransmitter receptor binding, in "Proteins of the Nervous System - Structure and Function", B. Haber, J. R., Perez-Polo and J. D. Coulter, eds., Alan R. Liss Inc., New York.

Kendall, D. A., Taylor, D. P., and Enna, S. J., 1983, ^{3}H-Tetrahydrotrazodone binding: association with serotonin$_1$ binding sites. Mol. Pharmacol., in press.

Lang, B., Barnard, E. A., Chang, L. R., and Dolly, J. O., 1979, Putative benzodiazepine receptor: a protein solubilized from brain. FEBS Lett., 104:149-153.

Miller, R. J. and Hiley, C. R., 1974, Anti-muscarinic properties of neuroleptics and drug-induced parkinsonism. Nature (London), 248:596.

Möhler, H., Polc, P., Cumin, R., Pieri, L., and Kettler, R., 1979, Nicotinamide, a brain constituent with benzodiazepine-like actions. Nature, 278:563.

Pert, C., and Snyder, S. H., 1974, Opiate receptor binding of agonists and antagonists affected differentially by sodium. Mol. Pharmacol., 10:868.

Peroutka, S. J., and Snyder, S. H., 1981, Interactions of antidepressants with neurotransmitter receptor sites, in "Antidepressants: Neurochemical, Behavioral and Clinical Perspectives", S. J. Enna, J. B. Malick, and E. Richelson, eds., Raven Press, New York.

Peroutka, S. J., U'Prichard, D. C., Greenberg, D. A., and Snyder, S. H., 1977, Neuroleptic drug interactions with norepinephrine α-receptor binding sites in rat brain. Neuropharmacology, 16:549.

Reisine, T.D., Fields, J. Z., Stern, L. Z., Johnson, P.C., Bird, E. D., and Yamamura, H. I., 1977a, Alterations in dopaminergic receptors in Huntington's disease. Life Sci. 21:1123.

Reisine, T. D., Fields, J. Z., Yamamura, H. I., Bird, E. D., Spokes, E., Schreiner, P. S. and Enna, S. J., 1977b, Neurotransmitter receptor alterations in Parkinson's disease. Life Sci., 21:335.

Rosenthal, H. E., 1967, Graphic method for the determination and presentation of binding parameters in a complex system. Anal. Biochem., 20:525.

Scatchard, G., 1949, The attraction of protein for small molecules and ions. Ann. N.Y. Acad. Sci., 51:660.

Schulster, D., and Levitzki, A., 1980, "Cellular Receptors for Hormones and Neurotransmitters", John Wiley & Sons, New York.

Shibuya, J., Bowie, D., and Pert, C., 1977, Opiate receptor ligand in cerebrospinal fluid: a simple radioreceptor assay requiring no preliminary purification. Proc. Soc. Neurosci., 3:460.

Simantov, R., Childers, S., and Snyder, S. H., 1977, Opioid pep-
    tides: differentiation by radioimmunoassay and radioreceptor
    assay, Brain Res., 135:358.
Sulser, F., and Vetulani, J., 1977, The functional post-synaptic
    norepinephrine receptor system and its modification by drugs
    which either alleviate or precipitate depression, in "Models
    in Psychiatry and Neurology", E. Usdin and I. Hanin, eds.,
    Pergamon Press, New York.
Wastek, G. J., Stern, L. Z., Johnson, P. C., and Yamamura, H. I.,
    1976, Huntington's disease: regional alterations in mus-
    carinic cholinergic receptor binding in human brain. Life
    Sci., 19:1033.
Yamamura, H. I., Enna, S. J. and Kuhar, M. J., 1978, Neuro-
    transmitter Receptor Binding", Raven Press, New York.

BINDING LIGANDS TO RECEPTORS:

ADDITIONAL COMMENTS

Alexander Levitzki

Department of Biological Chemistry
The Hebrew University of Jerusalem
91904 Jerusalem, Israel

Radioligand receptor assays enable the investigators to study directly the number of receptors in a tissue and the type of binding displayed by different ligands. In this chapter we shall address mainly the question what can we learn about receptors and their mode of action from ligand binding studies.

1. *Biochemical characterization of receptors*

Receptors have gained an interdisciplinary status in the life sciences and the medical sciences. Membrane receptors mediate the action of hormones and neurotransmitters, and therefore function as essential mediators of information transfer. The molecular mechanisms that underlie the action of receptors vary but one can, in fact, identify a few "families" of receptors which are typified by their biochemical uniqueness (Table I). The biochemical classification outlined in Table I demonstrates that there is a limited number of biochemical mechanisms which are the basis of the action of numerous receptors. Clearly, this type of classification is of limited value to the pharmacologist who classifies receptors according to their ligand specificity.

2. *Time scale of receptor action*

The onset of the biochemical event triggered by the receptor occupancy occurs with a certain delay of time, varying from receptor to receptor. For example, very "fast" receptors such as the nicotinic receptor, are activated within a few milliseconds. Conversely, slower receptors such as the muscarinic receptor, are activated within 100-300 milliseconds after the ligand occupancy step. Other receptors may be even "slower" in their response time. For

Table I. A Biochemical Classification of Membrane Receptors

Type of Biochemical Signal	Example of Receptors
A. Ion flux	
(a) influx of Na^+ (depolarization)	nicotinic receptor
(b) influx of Ca^{2+}	muscarinic (type 1), α_1--adrenergic, H1 histaminergic
(c) influx of Cl^-	GABAergic, glycine
B. Activation of adenylate cyclase	glucagon, ACTH, β-adrenergic, vasointestinal peptide (VIP), adenosine (A2)
C. Inhibition of adenylate cyclase	muscarinic (type 2), α_2--adrenergic, adenosine (A1)
D. Phosphorylation of a receptor or a receptor-linked component	EGF, insulin

example, the adenylate cyclase system is activated within 2-4 sec., subsequent to agonist occupancy. The time scale of receptor-linked events differs among receptors also at the onset of the process of desensitization. For example, the desensitization of the nicotinic receptor occurs within 1 sec., whereas the uncoupling of the β-receptor from the cyclase system occurs within a few minutes or longer. Other receptors such as the insulin receptor or the EGF receptor are "down regulated" and internalized. The presence of such a process makes the quantity of receptors on the membrane a variable quantity. It must be emphasized that the number of receptors may be variable in other systems too. Thus, for example, β-adrenergic receptor or α_2-adrenergic receptor down-regulation or up-regulation is a well established phenomenon. These few examples demonstrate the possible variability in the number of receptors on the membrane, according to the physiological state of the tissue from which the membrane was prepared.

3. Defining a receptor

Since the introduction of radioactively labeled drugs, the radioligand binding assay has become a very popular type of activity of the molecular pharmacologist. Evidently, not very high affinity site for a radioactive compound is a "receptor". For a high affinity site of a radioactively labeled compound to qualify as receptor, more is required than just the ability to obtain a proper Scatchard plot. (It has once been asked: "If I step on horse manure and some of it sticks to my sandals, does it necessarily mean that my sandals possess receptors for horse manure?!") One must demonstrate that (a) displacement of the ligand from its high affinity binding site by other drugs follows a well defined potency series typical of the receptor, and (b) binding of the particular ligand under study correlates with some biochemical activity elicited or inhibited by that drug. Such activity is, for example, activation or inhibition of the flux of a certain ion, the activation or inhibition of cAMP or cGMP formation, constricting or relaxing a muscle preparation, etc.

4. *Measuring binding*

Numerous methods have been devised in order to quantitate the number of receptors in a tissue preparation, a membrane preparation or even on intact cells. A detailed discussion on the advantages and disadvantages of the different methods has been given elsewhere (1); here with shall summarize these points in Table II. Obviously, the method of filtration has gained much popularity because of its reliability and simplicity. Indeed, whenever the life time of the ligand-receptor complex RL is long, it is advisable to use this method. One should, however, add a word of caution: not in every case when the dissociation constant is high, is the life time of the complex short. For example, suppose the K_D for a certain RL is 1×10^{-8} M; this could result from k_{on} of 1×10^8 M^{-1} sec^{-1} (rate constant of ligand binding is diffusion controlled) and $k_{off} = 1.0$ sec^{-1} ($t_{1/2} = 0.69$ sec). It is, however, possible that k_{on} is 1×10^7 M^{-1} sec^{-1} and $k_{off} = 0.1$ sec^{-1} ($t_{1/2} = 6.9$ sec). In the former case, the use of the filtration method is out of the question, whereas in the latter case it is borderline. In such a case, lowering the temperature may slow down k_{off} (and k_{on}), bringing the $t_{1/2}$ into the range of feasibility for a filtration experiment.

5. *What can we learn from binding experiments?*

One can learn a great deal from ligand binding experiments, and in this section we shall bring a few examples. One must be aware, however, that in order to draw the conclusions which are relevant to the mechanism of receptor action, one must perform the receptor binding experiments under the same conditions under which the biochemical and physiological experiments are carried out. For example, the drug ^3H-clonidine (or ^3H-paraamino-clonidine) binds with very high affinity to α_2-adrenergic receptors in the absence of guanyl nucleotides and of Na^+ as well at $4°C$, but with very much diminished affinity at $37°C$ in the presence of 140 mM NaCl and 0.1 mM GTP, a situation which more closely mimicks the physiological environment of the receptor. Numerous other examples, where the behavior of the receptor is strongly altered by the environment, exist. When a set of binding data and displacement experiments is

available, what can we learn from it?   A great deal, if we relate
it to the physiological situation.

Table 2. Properties of the Different Binding Techniques

Type of Method	Quantity Measured	Advantages	Disadvantages
Filtration	RL, L_{free}	Fast, many replicates	Cannot be performed when $t_{1/2}$ of the receptor-ligand complex is less than ∿15 sec
Ultracentri[a] fugation	RL, L_{free}	May be used when $t_{1/2}$ of the RL complex is small, namely, when ultrafiltration cannot be used	Slow, few replicates, high non-specific binding sometimes
Equilibrium dialysis	RL + L_{free}, L_{free}	True equilibrium method	Slow, cumbersome
Flow dialysis[b] (Colowick and Womack's method)	L_{free}	Fast, many points from the <u>same</u> sample	
Gel-filtration	RL	Suitable for soluble receptors	Quite slow
Polyethylene[c] -glycol (PEG) precipitation	RL, L_{free}	Suitable for soluble receptors; assay must be performed either by centrifugation or by filtration	Much more rapid than column assays

[a]Since in this method one deals with a pellet that is more difficult
to wash than a precipitate on a filter, the included solvent contributes significantly to the increased "non-specific" binding.
[b]This method is described in Ref. 2.   [c]See Ref. 3.

     Another issue is that of cooperativity. One can determine
whether the process of binding is cooperative and correlate its
pattern with the pattern of the response. For example, in the case
of the nicotinic receptor, it has been recognized a long time ago
that the *response* is cooperative. Thus, a direct binding experiment
may prove useful, as one can elucidate whether the cooperativity is
at the level of agonist binding or results from the peculiarities of
the signaling system. Unfortunately, and because of very good reasons (4), in this particular instance this issue has not as yet been

resolved. Another example is that of the binding of ^{125}I-insulin which was found to yield non-Michaelian isotherms with the appearance of *negative cooperativity*. Kinetic experiments have corroborated the assertion of De-Meyts *et al.* that the curvilinear Scatchard plots reveal negative cooperativity rather than a heterogeneous population of sites. The availability of non-cooperative insulin analogs allow the examination of ^{125}I-insulin binding in the presence of these non-cooperative analogs. Such experiments may establish the nature of the cooperativity in the binding (5,6). However, it should be emphasized that the exact molecular mechanism of the negative cooperativity cannot be deciphered exclusively on the basis of binding studies.

Binding studies can also help to understand the phenomenon of spare receptors. If the binding isotherm can be compared with the dose response curve, one may calculate exactly (1,4) the "excess" of receptors over those needed to generate the maximum physiological response. These and other issues are discussed in detail in Ref. 1 and in a recent monograph (4).

References

1. A. Levitzki, Quantitative aspects of ligand binding to receptors, *in*: "Cellular Receptors for Hormones and Neurotransmitters," D. Schulster and A. Levitzki, eds., Wiley & Sons, pp. 9-28,1980.
2. S.P. Colowick and F.C. Womack (1969) Binding of diffusable molecules by macromolecules: Rapid measurement by rate dialysis, *J. Biol. Chem.* 244:774.
3. B. Desbuquois and G. Aurbach (1971) Use of polyethylene glycol to separate free and antibody-bound peptide hormones in radioimmunoassays, *J. Clin. Endocrinol.* 33:732.
4. A. Levitzki (1983) *in*: "Receptor Binding and Receptor-to-Signal Coupling", in preparation.
5. Y.I. Henis and A. Levitzki (1979) Ligand competition curves as diagnostic tool for delineating the nature of site-site interactions: Theory, *Eur. J. Biochem.* 102:449.
6. Y.I. Henis and A. Levitzki (1980) The sequential nature of the negative cooperativity in rabbit muscle glyceraldehyde-3-phosphate dehydrogenase, *Eur. J. Biochem.* 115:59.

BASIC CONCEPTS TO ANALYZE BINDING DATA USING PERSONAL COMPUTERS: The "RECEPT" program

Fabio Benfenati and Vincenzo Guardabasso

Institute of Human Physiology, University of
Modena, via Campi 287, I 41100 Modena, Italy and
"Mario Negri" Institute for Pharmacological
Research, Milan, Italy

Microcomputers have become more easily available in
recent years thanks to improvements in technology and lower
prices. Many biological research laboratories now have a
small computer available for data analysis, enabling their
staff to avoid long, tedious and uncertain calculations
with the sole help of a pocket calculator, and overcoming
the long wait for quantitative results when calculations
have to wait until all planned experiments have been
completed. Connection to a large computing centre, once
the only way to achieve computerized data analysis, is not
always convenient; it takes too long to get acquainted
with large-scale system working procedures, usually not
user-friendly, and the rates are high.
       Well recognized advantages of using a computer are
the following:
- data can be analyzed in a few minutes after each experiment or on a day-to-day basis. The immediacy of quantitative analysis is very useful as it permits rearrange-
ments of future experimental work in the short term;
- computerized data analysis is fast, the time taken being
virtually only the time necessary for data entry;
- quantitative data analysis is much more reliable and
objective than qualitative or graphic estimates. There are
other advantages too, whose relevance is sometimes overlooked. For example, most small computers give graphic
facilities so data can be quickly arranged in graphs,

plots or histograms on the screen before further processing. This helps to give an immediate idea of experimental outcomes, and each step of data analysis can be followed to check the calculations. Graphs are handy tools
for evaluation of experiments besides being a form in
which to present results in papers. Moreover, a computer
with a plotting device can produce high-quality drawings,
ready for preparing slides or illustrations. Then too,
data can be easily stored on magnetic media (usually
inexpensive flexible floppy-magnetic disks) as data files.
Once stored, data can be referred to simply as a name, and
can be easily retrieved, modified or processed many times
without further input.

Computers however have some limitations and some
caution is needed for rational use of Electronic Data
Processing. A computer is neither a clever mathematician
nor an infallible "black box". It is merely a device which
performs operations the programmer has planned for it.
For example, it cannot recognize a "foolish" number yielded
by an empty vial erroneously put in a scintillation
counter, unless it has been programmed to do so. Therefore
the researcher who uses a computer must check data before
they are fed in or before they are processed and must know
the steps the programs follow and how they were designed,
if he is to understand fully what the machine "feeds out",
and to be sure of the meaning and consistency of results.
The following section describes a group of programs for
binding assays to show in detail how data can be processed
and to give an explanatory outline to anyone interested
in writing a program for quantitative analysis of data
from radioligand binding experiments.

The RECEPT programs were prepared for quantitative
analysis of ligand-receptor interactions at equilibrium
and can be divided into two parts, one for saturation
studies and the other for inhibition (or displacement)
experiments. Despite this division both programs have a
similar structure and their main stages develop as shown
in Fig. 1, described below:
- DATA INPUT : read data from keyboard or data file;
- DATA REDUCTION (A.) with the options of data printout,
correction, recording in data file and plot;
- LINEAR TRANSFORMATIONS (B.) followed by linear regression, fitting evaluation, printout of the results and plot;

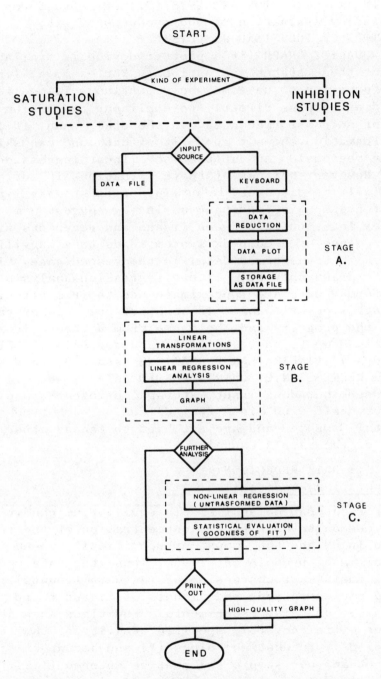

Fig. 1 - Flow chart of the RECEPT program.

-NONLINEAR REGRESSION ANALYSIS (C.) of untrasformed data
followed by evaluation of the goodness of fit;
-PRINTOUT OF THE FINAL RESULTS;
-HIGH-QUALITY GRAPH (if a plotter device is available).
Some of these stages are optional. For example, linear
transformation of data may be superfluous if one is
intending to pass directly to nonlinear regression analysis of untransformed data. On the other hand, if linear
transformation shows a good fit of data one can skip nonlinear regression and stop at the linear regression analysis. However, neither of these situations is advisable,
as it will be discussed later. In the first case linear
transformation is important as it may suggest a more
complex interaction between ligand and receptors since
the presence of a complex system leads to a curvilinear
pattern in transformed data. In the second case, final
results drawn only from linear regression analysis of
transformed data are approximate and their statistical
meaning is questionable. These brief preliminary remarks
about the programs and their use from a theoretical stand-
point lead up to the detailed features. DATA REDUCTION
(stage A.), LINEAR TRANSFORMATIONS (stage B.) and NON-
LINEAR REGRESSION ON UNTRANSFORMED DATA (stage C.) are
discussed, but data input, storage, printout and plotting
are not dealt with since they differ on each type of
Personal Computer and are not hard to manage after a little
study of the user's manual.

STAGE A.: DATA REDUCTION
1. Saturation studies (Table 1A).

     For saturation analysis in equilibrium conditions
it is common to express the concentration of the free
ligand in nM and the amount of specifically bound ligand
in pmol/mg of protein or tissue. Thus, the data reduction
stage consists of conversion of raw data (counts) into
ligand concentrations and amounts of ligand bound to membranes. To do this, experimental conditions such as
counter efficiency (E), specific activity of the labeled
ligand (SA), incubation volume (V) and amount of tissue
(or protein) per sample (W) must be entered (1,2). The
SA, usually in Ci/mmol (or nCi/pmol), is transformed to
cpm/pmol (CSA or converted SA) by means of E and dpm/nCi
conversion factor (2220; Eq.1). Furthermore when the

labeled ligand is carried by increasing amounts of unlabeled ligand,CSA must be further corrected for each concentration with the labeled/total ligand ratio (eq.2).
Then for each experimental point(each ligand concentration) the counts of total ligand added to tubes (TA or
total ligand activity), total bound ligand (TB) and
unspecifically bound ligand in the presence of unlabeled
displacer (UB) are entered; from these values the free
ligand concentration (F in nM) and specifically bound
ligand (B in pmol/mg of protein) are easily calculated
according to the eq.3 and 4. It is assumed that counter
data entered are the mean of at least duplicate or triplicate determinations. To simplify the description, standard deviations of mean values are discarded, and variances
are assumed to be approximately constant over the whole
range of concentrations tested. Unfortunately this is only
an approximation. Weighting of data according to their
variance is desirable. Alternatively, when not using data
transformations, each single counter determination can be
processed as individual data.

In the program, experimental conditions are located
in simple variables; the SA must be put in an array only
if it is not constant. Experimental data are always
located in arrays. An array (a vector or a matrix) is a
table of numbers; it is used to store an ordered set of
numbers identifiable by their position in the array. An
array has a name (like a variable) and a subscript (in
brackets) which identifies the array position occupied
by a given number. Arrays are easily used in FOR/NEXT
loops, using the index variable of the loop as array
subscript in INPUT-OUTPUT or data conversions. Here an
example is shown, for allocation of Bound and Free values
for N experimental points in the arrays B() and F() following the entry of TA, TB and UB counts:

```
100  REM  ENTRY OF EXPERIMENTAL DATA AND B-F CALCULATIONS
110  FOR I = 1 TO N
120  PRINT "ENTER TOTAL ACTIVITY,TOTAL BOUND AND UNSPECIFIC BOUND LIGAND
     OF POINT ";I
130  INPUT TA,TB,UB
140  F(I) = (TA - TB) / CSA / V
150  B(I) = (TB - UB) / CSA / W
160  NEXT I
```

At the end of Data reduction, a plot of data is displayed
(X=Free, Y=Bound; Saturation Curve plot). Data should

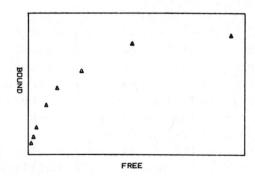

Fig. 2 - Saturation curve of 3H-spiperone binding to rat
         strital membranes.

show a curvilinear pattern (a rectangular hyperbola in
the simple case) (Fig.2).

2. Inhibition studies (Table 2A).

   In inhibition studies a fraction of binding sites is
occupied using a fixed concentration of labeled ligand.
This system is tested with various concentrations of unlabeled ligand (inhibitor), and a set of (specific or total)
"residual binding" values for the labeled ligand is
obtained (2). The specific (or total) binding of labeled
ligand in the absence of inhibitor is taken as "control
binding" (or 100% binding). The ratio "residual binding"/
"control binding" at each concentration of inhibitor,
multiplied by 100, gives the"percentage of control" residual binding at that concentration. At each concentration
of inhibitor: "percentage inhibition" = 100-"percentage
residual binding". If the definition of unspecific binding
for that labeled ligand is known, it is advisable to
calculate inhibition data as a percentage of specific
binding. However, when the labeled ligand binds to different receptors (as in the case of 3H-spiperone, 3) and
the various components have to be differentiated using
selective inhibitors, data should be expressed as percentage of total binding. For further descriptions we assume
measuring an unspecific binding value (UB) (e.g. in presence of both labeled ligand and an excess of the same
unlabeled ligand) and go on with data expressed as
a percentage of specific binding. Among the experimental
conditions, incubation volume, specific activity of the
ligand and total ligand activity are used to determine

TABLE 1.   SATURATION STUDIES

STAGE A. : DATA REDUCTION

Entry of experimental conditions :

- N = number of ligand concentrations tested
- V = incubation volume (ml)
- SA = specific activity of the ligand (Ci/mmol=nCi/pmol)
- W = protein or tissue content per tube (mg)
- E = counter efficiency = (cpm/dpm)*100

SA conversion :   - if SA is constant (at all concentrations) :

(1) $CSA(\frac{cpm}{pmol}) = SA(\frac{nCi}{pmol}) * 2220(\frac{dpm}{nCi}) * \frac{E}{100}(\frac{cpm}{dpm})$

- if SA is not constant, enter the hot/total ligand ratio (HTR) for each concentration; then, for each concentration :

(2) $CSA = CSA * HTR$

Entry of experimental data (counts) :

for each concentration of ligand   - TA = total ligand activity (amount added to each tube)
- TB = total bound ligand
- UB = unspecific bound ligand

Calculations :

(3) $F(nM) = \frac{\text{free ligand}}{\text{concentration}} = \frac{TA - TB}{CSA * V} (\frac{pmol}{ml} = \frac{nmol}{l})$.

(4) $B(\frac{pmol}{mg}) = \text{specific bound ligand} = \frac{TB - UB}{CSA * W}$

Data plot (B vs F)

(continued)

(according to Eq.11) the labeled ligand concentration (TLC) in nM at which the inhibition study has been performed. This is useful in case of competitive inhibition to convert IC50 values from experimental data into Ki values (or dissociation constant of the inhibitor-receptor complex) as follows:

$$Ki = IC50/(1+TLC/Kd)$$

(where Kd is the dissociation constant of labeled ligand), since IC50 differs at different labeled ligand concentrations (4). In the case of uncompetitive inhibition the IC50 value is independent of the ligand concentration and corresponds to Ki (IC50 = Ki).

Necessary experimental conditions are the total bound labeled ligand (TB) and the unspecifically bound labeled ligand (UB) in counts per tube. Then, after entering the experimental data for each point (i.e. the

TABLE 1.   SATURATION STUDIES (Continued)

STAGE B. : LINEAR TRANSFORMATIONS

SCATCHARD PLOT :   X = B
                   Y = B/F
                   after linear regression analysis

(5)   B/F = slope * B + Y-intercept

     Results: Kd (nM) = - 1/slope

              Bmax $(\frac{pmol}{mg})$ = X-intercept = $- \frac{Y-intercept}{slope}$

EADIE-HOFSTEE PLOT : X = B/F
                      Y = B
                      after linear regression analysis

(6)   B = slope * B/F + Y-intercept

     Results: Kd (nM) = - slope

              Bmax $(\frac{pmol}{mg})$ = Y-intercept

HILL PLOT :   X = log F
              Y = $\log (\frac{B}{Bmax - B})$
              after linear regression analysis

(7) $\log (\frac{B}{Bmax - B})$ = slope * log F + Y-intercept

     Results: Hill Coefficient = slope

              log Kd = X-intercept = $- \frac{Y-intercept}{slope}$

(continued on page 51)

---

inhibitor concentration I in Molarity and the residual bound labeled ligand in counts) the percentage of residual total (or specific) binding are calculated (eq.12 and 13). The percentage of inhibition of specific (or total) binding is then easily obtained (eq.14);it represents the amount of inhibitor bound to receptors, hence the "effect" of increasing concentrations of inhibitor. At the end of calculations a plot of data is displayed (X = log (I) Y =% RB: inhibition curve plot). Data should show a sigmoidal, fairly symmetrical shape (Fig. 3).

STAGE B. : LINEAR TRANSFORMATIONS
1. Saturation studies (Table 1B).

The linear transformations mostly used for saturation studies are the Scatchard and Eadie-Hofstee analyses (eq. 5,6) (for critical review see 1,2,5,6,7). After transformation of experimental data as shown in Table 1B, linear regression analysis is used to find the "best fitted" line

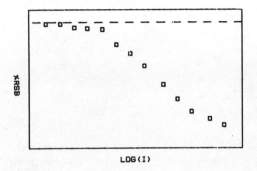

Fig. 3 - Inhibition curve of 3H-spiperone binding by
         increasing concentrations of bromocriptine.

between points (see appendix A). The binding parameters
Kd (dissociation constant of the ligand-receptor complex
at equilibrium) and Bmax (maximal amount of ligand bound
per weight unit) are then computed from the parameters of
the line (slope and Y-intercept). However, application
of standard linear regression analysis to this kind of
transformed data is not statistically correct (6,7). The
requirements of theory are not fulfilled (errors in Y are
not normally distributed and not independent of errors in
X); calculated values for parameters and their standard
deviations are not statistically meaningful. Thus binding
parameters obtained from linear transformations are approximate and a formally more correct analysis must be done
on untrasformed data (see below and 6,7,8). Another point
is that in Scatchard or Eadie-Hofstee plots a point with
large experimental error may lie far from the other well
aligned points, greatly influencing parameter determination. If this is the case, and if the corresponding point
is clearly out of the saturation curve (untransformed
data) it can be excluded before linear regression analysis.
Objective criteria exist for rejecting "outliers" (study
of the confidence intervals and outliers test) and tell
which points are statistically suspect (17). The following BASIC subroutine enables the user to remove from
arrays X() and Y() (containing N points) the coordinates
of points:

```
3000  REM    SUBROUTINE TO DELETE POINTS
3010  K = O
3020  PRINT "HOW MANY POINTS ARE YOU GOING TO DELETE "
3030  INPUT M
3040  FOR L = 1 TO M
```

```
3050  PRINT "WHICH POINT"
3060  INPUT Z
3070  Z = Z - K
3080  J = 0
3090  FOR I = 1 TO N
3100   IF I = Z THEN 3140
3110  J = J + 1
3120  X(J) = X(I)
3130  Y(J) = Y(I)
3140   NEXT I
3150  K = K + 1
3160   NEXT L
3170  N = N - M
3180  RETURN
```

A very important step is the assessment of linearity (see also appendix A). If experimental points in the Scatchard ot Hofstee plot are randomly distributed in a linear pattern along the regression line (Fig. 4A), then the hypothesis of a simple binding interaction, with one class of independent binding sites, is confirmed. If this is the case, the binding parameters obtained from linear regression on these plots are a good approximation of "true" binding parameters (9). If however a curvilinear pattern is present (Fig. 4B), a more complex interaction is likely

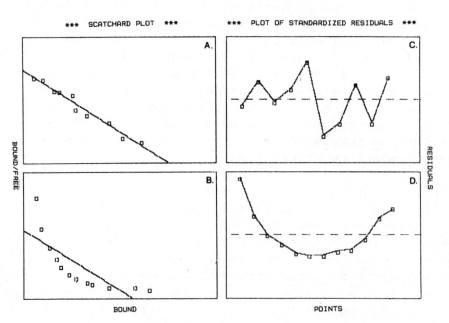

Fig. 4 - A. Scatchard plot of 3H-spiperone binding to rat cerebral cortex; B. Scatchard plot of 3H-clonidine binding to rat hypothalamus; C.Plot of stand. residuals of A.; D. Plot of stand. residual of B.

TABLE 1.   SATURATION STUDIES (Continued)

STAGE C. :   NON-LINEAR REGRESSION

UNTRANSFORMED DATA :   X = F (nM)
                        Y = B (pmol/mg)

MODEL FOR SIMPLE_CASE

(8) $B = \dfrac{P_1 * F}{P_2 + F}$

            Results: P_1 = Bmax (pmol/mg)
                       P_2 = Kd (nM)

MODEL FOR MULTIPLE_INDIPENDENT_SITES

(9) $B = \sum\limits_{i=1}^{n} \dfrac{P_{1i} * F}{P_{2i} + F}$

            with i = number of non-interacting sites

            Results: P_{1i} = Bmax (pmol/mg) of site i
                       P_{2i} = Kd (nM) of site i

MODEL FOR BIVALENT_COOPERATIVITY

(10) $B = \dfrac{\dfrac{P_1 * F}{P_2} * \left(1 + \dfrac{P_3 * F}{P_2}\right)}{1 + 2 * \dfrac{F}{P_2} + P_3 * \left(\dfrac{F}{P_2}\right)^2}$

            Results: P_1 = Bmax (pmol/mg)
                       P_2 = initial Kd (nM)
                       P_3 = interaction factor $\left(\dfrac{\text{initial Kd}}{\text{final Kd}}\right)$
                               if = 1 ⟶ no cooperativity
                               if > 1 ⟶ positive cooperativity
                               if < 1 ⟶ negative cooperativity
---

(10). In this case binding parameters obtained from linear
regression would be meaningless and misleading, and a more
complex analysis is needed (6,8,10,11).

Another tool to check whether the "simple model"
hypothesis holds for the data studied is the Hill plot
(1,2) (see Table 1B, eq.7). It calls for a previous
determination of Bmax. The Hill coefficient n (the slope
of regression line) should be 1 in a simple bimolecular
ligand-receptor interaction; n < 1 suggests the presence
of multiple binding sites or negative cooperativity and
n > 1 suggests positive cooperativity. These hypotheses
can be verified and compared by analysis of untransformed

data (see below). Due to the tendency of points in a Hill
plot to deviate from linearity at the extremes, only points
whose B values lie between 10% and 90% of the Bmax should
be used for linear regression (2). The remaining points
can be then deleted using the BASIC subroutine described
above.

2. Inhibition studies (Table 2B).

One common method of transforming an inhibition curve
into a line is the Log-Probit plot (eq.15) in which the
calculated binding percentage (usually %RSB) is converted
into probits or probability units (12). Probit values for
given percentages are found in statistical tables and can
be put into a BASIC program using DATA statements. The
x-value corresponding to Probit 5 is log(IC50), that is
the logarithm of the concentration of unlabeled ligand
that causes 50% inhibition of labeled drug binding to a
specific site.

Here is an example of the use of DATA statements in
a BASIC program:

```
2000   REM  FIND CORRESPONDING PROBIT VALUE FOR PERCENT RESIDUAL BINDING

2010   FOR I = 1 TO N
2020   IF (Y(I) < 1) OR (Y(I) > 99) THEN 2120
2030   REM  ROUND OFF THE PERCENT RESIDUAL BINDING
2040 Y(I) =  INT (Y(I) + .5)
2050   REM  SET DATA READING AT THE BEGINNING AND START TO SCAN DATA VALUES

2060   RESTORE
2070   FOR K = 1 TO Y(I)
2080   READ PROBIT
2090   NEXT K
2100   REM  PUT THE FOUND PROBIT VALUE IN ARRAY P()
2110   P(I) = PROBIT
2120   NEXT I
2130   REM
2140   REM
2150   REM
10000  DATA   (PROBIT VALUE AT 1%), (AT 2%), (AT 3%),...............
10010  DATA   .........
10020  DATA   .........
```

For inhibition curves too the Hill coefficient can
be determined from a Logit-Log plot (or indirect Hill
plot) (2) as shown in eq. 16. The Hill coefficient is
approximated, because correct application of Hill's equation requires concentration of Free inhibitor, substituted
here by total concentrations. For both Log-Probit and
Logit-Log plots, points whose %ISB lie between 0-10% and
90-100% should not be considered for linear regression
analysis since at these levels the plots deviate from
linearity.

Eadie-Hofstee and Scatchard plots can be used in inhibition studies too. The inhibitor concentrations are an approximation of the Free values, %ISB values are a measure of the inhibitor specifically bound to receptors and IC50 corresponds to the Kd value in saturation experiments. However Bmax, the maximum number of receptors that could be occupied by the inhibitor, has been set at 100% in DATA REDUCTION stage and cannot be determined. In inhibition studies too, it is very important to check linearity in Eadie-Hofstee and Scatchard plots in order to detect more complex interactions between the unlabeled ligand and receptors (10,13).

STAGE C: NONLINEAR REGRESSION ON UNTRASFORMED DATA

Linear transformations are simple and useful for checking that experimental data follow the linear "simple case" model (one ligand and one homogenous class of non-cooperative binding sites) but they can give approximate and sometimes biased results (see above). A statistically more reccomendable approach is to calculate binding parameters directly from untransformed data by finding the parameters of equations describing the saturation curve (binding isotherm) by means of nonlinear least-squares regression analysis (Appendix A) (6,7,8). This mathematical method analyzes data according to different binding models (e.g.: one saturable site, multiple independent sites, site-site cooperativity), defining the corresponding equations in the BASIC program in a subroutine or DEF FN statement. Our programs use a version of NL-FIT (14), tailored for binding studies by adding a set of equations for the different binding models used.

1. Saturation studies (Table 1C)

The equations used for nonlinear regression analysis of untrasformed Free and Bound data are shown in Table 1C. Eq.8 and 9 are derived from Feldman (15) for one ligand interacting with one or multiple classes of independent binding sites. Eq.10 is used in the case of one class of sites with bimolecular cooperativity (9). A printout of saturation study results is shown in Fig. 5. After data analysis "goodness of fit" must be assessed and various equations according to different hypotheses must be compared (see Appendix A).

```
****************************************************
**    NON-LINEAR FITTING ANALYSIS OF BINDING DATA    **
****************************************************

EXPERIMENT : 3H-SPIPERONE BINDING - STRIATUM
MODEL : ONE-SITE SATURATION CURVE

N              X              Y

1              .014522        .032567
2              .026577        .048953
3              .040387        .077525
4              .086429        .140781
5              .137497        .187289
6              .249199        .236251
7              .483639        .313809
8              .937813        .336768

WEIGHT = 1/Y^2

N.             INITIAL PARAMETER

1              .4
2              .2

)))))))))) ((((((((((
))))) CONVERGENCE (((((
)))))))))) ((((((((((

NO. OF ITERATIONS = 3
     SIGMA(SQ.ERR.) = 8.405200  E-05
     UNWEIGHTED SIGMA(SQ.ERR.) = 3.946783  E-04
     VARIANCE = 1.400866  E-05
     STD.DEV. = 3.742815  E-03
     R^2 = .994933.
     LAST LAMBDA = 1000


FIN. PARAM.     STD.DEV.        CV%
1 .415610       .018944         4.558144
2 .179391       .012653         7.053598


**** BINDING SITE :
KD = .179391     NM
BMX = .415610    PMOL/MG PROT

*** SATURATION PLOT ***
```

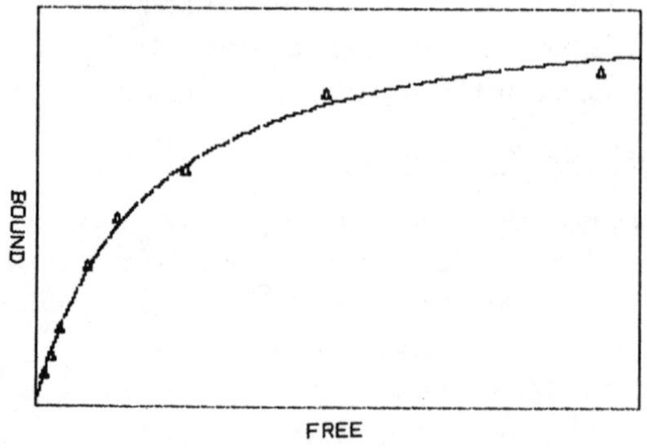

Fig. 5 - Results printout of 3H-spiperone saturation curve in rat striatum analyzed by nonlinear regression using Eq.8.

TABLE 2.   INHIBITION STUDIES

STAGE A. : DATA REDUCTION ────────────────────────

The effect of inhibitor is expressed as   % of control binding of labeled ligand. "Control binding" can be :
- total bound labeled ligand without inhibitor = TB
- specific bound labeled ligand without inhibitor = SB

Entry of experimental conditions :

- N = number of inhibitor concentrations tested
- V = incubation volume (ml)
- SA = specific activity of the ligand (Ci/mmol=nCi/pmol)
- E = counter efficiency = (cpm/dpm)*100
- TA = total ligand activity (amount added to each tube)
- TB = total bound labeled ligand
- UB = unspecific bound labeled ligand (if results are expressed as %TB then UB = 0)

SA conversion : SA(Ci/mmol) ⟶ CSA(cpm/pmol) (see table 1)

Calculation of total ligand concentration (TLC) :

$$(11) \qquad TLC(nM) = \frac{TA}{CSA * V} \ (\frac{pmol}{ml} = \frac{nmol}{l})$$

Entry of experimental data :

for each concentration tested - I = concentration of unlabeled ligand (inhibitor) (M)

- RB = residual bound labeled ligand (counts)

Calculations :

$$(12) \qquad \%RTB = \frac{percent\ residual}{total\ binding} = \frac{RB}{TB} * 100$$

$$(13) \qquad \%RSB = \frac{percent\ residual}{specific\ binding} = \frac{RB}{SB} * 100$$

$$\%ITB = \frac{percent\ inhibited}{total\ binding} = 100 - \%RTB$$

$$(14)$$

$$\%ISB = \frac{percent\ inhibited}{specific\ binding} = 100 - \%RSB$$

Data Plot (%RTB or %RSB vs log I)

(continued)

---

2. Inhibition studies (Table 2C)

Inhibition curves with untransformed data can be analyzed according to the hypotheses of one or multiple binding sites, or cooperativity. Appropriate equations can be derived from the general binding model (15) when two ligands interact with binding sites. Under some approximations (concentrations of Free labeled and unlabeled ligands are replaced by total concentrations) and the assumption that the labeled ligand has the same affinity

TABLE 2.   (Continued)

STAGE B. :   LINEAR TRANSFORMATIONS

LOG-PROBIT PLOT : $X = \log I$
$Y = $ probits(%RSB) taken from appropriate tables
after linear regression analysis

(15)   Probits = slope $*$ log I + Y-intercept

Result: $\underline{\log IC_{50}}$ = X-value corresponding to probit 5

LOGIT-LOG PLOT (indirect Hill plot) : $X = \log I$
$Y = \log (\frac{\%ISB}{100 - \%ISB})$

after linear regression analysis

(16) $\log (\frac{\%ISB}{100 - \%ISB})$ = slope $*$ log I + Y-intercept

Results: $\underline{Hill\ Coefficient}$ = slope (approximate)

$\underline{\log IC_{50}}$ = X-intercept = $- \frac{Y-intercept}{slope}$

EADIE-HOFSTEE PLOT : $X = \%ISB/I$
$Y = \%ISB$
after linear regression analysis

(17)   %ISB = slope $*$ $\frac{\%ISB}{I}$ + Y-intercept

Results: $\underline{IC_{50}}$ (M) = - slope

$\underline{Max\ Inhibition}$ (%) = Y-intercept

SCATCHARD PLOT : $X = \%ISB$
$Y = \%ISB/I$
after linear regression analysis

(18) $\frac{\%ISB}{I}$ = slope $*$ %ISB + Y-intercept

Results: $\underline{IC_{50}}$ (M) = - 1/slope

$\underline{Max\ Inhibition}$ (%) = X-intercept = $- \frac{Y-intercept}{slope}$

(continued)

(same Kd) for all sites even if the unlabeled ligand shows
different affinities (multiple sites), the complex equations can be simplified as in Eq.19 and 20. Should the
labeled ligand show different affinity for the sites, then
Eq.20 does not hold and the problem must be tackled through
careful experimental design and very complex analysis (8).
Alternatively inhibition curves can be fitted with the four
parameters logistic equation (Eq.21) which fits generic
sigmoidal dose-response curves (16). For inhibition studies
too, statistical evaluation of goodness of fit is needed
and various models tested must be compared (see appendix
A). Furthermore IC50 values have to be corrected for

TABLE 2. (Continued)

STAGE C. : NON-LINEAR REGRESSION

UNTRANSFORMED DATA : X = I (M)
                       Y = %ISB

MODEL FOR SIMPLE CASE

(19) $\%ISB = \dfrac{P_1 * I}{P_2 + I}$

         Results: P_1 = Max Inhibition (%)

                   P_2 = IC_{50} (M)

MODEL FOR MULTIPLE INDEPENDENT SITES

(20) $B = \displaystyle\sum_{i=1}^{n} \dfrac{P_{1i} * I}{P_{2i} + I}$

         with n = number of non-interacting sites

         Results P_{1i} = Max Inhibition (%) due to site i

                  $\%$ site i = $P_{1i} / \displaystyle\sum_{j=1}^{n} P_{1i}$

                  P_{2i} = IC_{50} for site i

LOGISTIC EQUATION

(21) $\dfrac{\%\ effect}{} = \dfrac{P_1 - P_4}{1 + (\dfrac{I}{P_3})^{P_2}} + P_4$

         where $\%$ effect = %RSB, %RTB, %ISB, %ITB

         Results P_1 = % of Min effect

                  P_4 = % of Max effect

                  P_3 = IC_{50} (M)

                  P_2 = slope (approximated Hill coefficient)

MODEL FOR BIVALENT COOPERATIVITY

         see eq.10 in Table 1.C using :
         - I in place of F
         - %ISB in place of B
         - % of Max inhibition in place of Bmax
         - IC_{50} in place of Kd

labeled ligand concentration according to the type of inhibition (see before and 4).

CONCLUSIONS

In this paper we have described "RECEPT", a group of programs to analyze experimental data from the most commonly studied ligand-receptor interactions at equilibrium. The programs have been developed for use on Personal

Computers, whose features (in terms of costs, speed and
precision) are adequate to deal with some relatively
complex problems too. They are currently used by researchers at our Institutes. Our purpose was to explain how to
design such programs by people with limited knowledge of
programming and algorithms. Only basic understanding of
Input-Output and Printer-Graphic statements is required
to implement at least stages A and B of the above elaboration procedures in BASIC language. Today data processing should not be considered a difficult step and a small
computer can save researchers time, while dealing quantitatively and objectively with experimental data analysis. It is however important to recall that a computer
is nothing but a device: mathematics and graphs serve for
evaluation of experimental results but do not give proofs.
Considerable experience is always needed in interpreting
experimental results.

HARDWARE AND SOFTWARE SPECIFICATIONS

The programs described here have been implemented in
BASIC language and run on Apple II (48 Kbytes memory) and
Hewlett-Packard HP-85 (32 Kbytes memory) computers. The
overall size of all programs and subroutines plus data
structures is about 100 Kbytes for HP85 and about 140
Kbytes for Apple II.

APPENDIX A: LEAST-SQUARES REGRESSION ANALYSIS

1. Linear regression

The general equation of a line is
$$Y = \text{slope} * X + Y\text{-intercept}$$
Linear regression analysis (17) is a standard statistical
method to find the numerical values of the two parameters
"slope" and "Y-intercept" for the best fitted line through
experimental points. The principle of "best fit" is that
the sum of squared differences between measured Y and calculated Y must be minimized ("least-squares method").
Linear regression also gives mean standard deviation of
regression, the coefficient of correlation r, the coefficient of determination r2 and standard errors of slope,
Y-intercept and X-intercept. The correlation coefficient
is a measure of correlation between variables X and Y

(perfect correlation when r-1 or r=-1. The determination coefficient, multiplied by 100, gives an estimate of the ssion; it is important to note that "good" values for these coefficients (e.g. r=0.99) do not mean that experimental points show a linear distribution.

The linearity of experimental points (correspondence to a line equation) can easily be checked looking at a graph of points and regression line. A random distribution of points, evenly scattered above and below but close to the line, is a good sign (Fig.4A). Any "pattern" (an ordered distribution of points: Fig.4B) should be viewed with suspicion as a likely indication of nonlinearity. Statistically more reliable methods are the Plot of Standardized residuals (17) and the Runs test (18). Residuals (or errors)for each given point X-value are the difference between the measured Y-coordinate and the Y-value calculated on the best fitted line (RES = Y-meas - Y-calc). The residuals can be standardized by dividing them by the standard deviation of regression. In the Plot of Standardized residuals points are plotted with X=variable X and Y=Std. res. Experimental points should lie randomly above (positive) and below (negative) the zero, with values between 2.0 and -2.0 (see Fig.4C,4D). Another analysis is the Runs test which studies the sequence of algebraic signs of residuals and compares the number of sign changes along the row of numbers with tables of theoretical distribution of random events (18).

Linear regression analysis can be performed on many hand-held calculators; it can be implemented in BASIC language as follows, assuming that the transformed coordinates of N experimental points are stored in arrays X() and Y():

```
4000  REM  LINEAR REGRESSION ANALYSIS SUBROUTINE
4010  REM  SET SUM VARIABLES TO ZERO
4020 J = 0
4030 K = 0
4040 L = 0
4050 M = 0
4060 R = 0
4070  REM  ACCUMULATE INTERMEDIATE SUMS
4080  FOR I = 1 TO N
4090 J = J + X(I)
4100 K = K + Y(I)
4110 L = L + X(I) ^ 2
4120 M = M + Y(I) ^ 2
4130 R = R + X(I) * Y(I)
4140  NEXT I
```

```
4150  REM  COMPUTE REGRESSION ANALYSIS
4160  REM   RO = DEVIANCE OF X : R1 = DEVIANCE OF Y : R2 = CODEVIANCE
4170  RO = L - J * J / N
4180  R1 = M - K * K / N
4190  R2 = R - J * K / N
4200  R3 = R2 /  SQR (RO * R1)
4210  R4 = (R1 - R2 * R2 / RO) / (N - 2)
4220  V = R2 / RO
4230  Z = K / N - V * J / N
4240  PRINT "LINE EQUATION : Y = SLOPE*X + Y-INTERCEPT"
4250  PRINT "N. OF POINTS = ";N
4260  PRINT "SLOPE = ";V
4270  PRINT "Y-INTERCEPT = ";Z
4280  PRINT "X-INTERCEPT = ": - Z / V
4290  REM  STATISTICAL EVALUATIONS
4300  PRINT "COEFFICIENT OF CORRELATION (R) = ";R3
4310  PRINT "COEFFICIENT OF DETERMINATION (R^2) = ";R3 * R3
4320  PRINT "VARIANCE OF REGRESSION (S^2) = ";R4
4330  PRINT "STD.DEV. OF REGRESSION (S) = "; SQR (R4)
4340  PRINT "STD.ERR. OF SLOPE = "; SQR (R4 / RO)
4350  PRINT "STD.ERR. OF Y-INTERCEPT = "; SQR (R4 * (1 / N + (J / N) ^ 2 /
      RO))
4360  PRINT "STD.ERR. OF X-INTERCEPT = "; SQR (R4 / V / V * (1 / N + (K /
      N / V) ^ 2 / RO))
4370  RETURN
```

2. Nonlinear regression

    Nonlinear regression analysis is a mathematical
method to find the numerical values of parameters of the
"best fitted" curve through points, when the equation of
the curve is known and is nonlinear in the parameters (19).
Once programs for this computational method were only available on large computers (8,20), but now many programs of
nonlinear regression running on Personal Computers have
been published (14,21,22). Programs for nonlinear regression starting an iterative search from initial estimates
of parameters, find the values that give the "best fitted"
curve between experimental points with their approximate
standard deviations.
    The advantages of this method are:
(i) it works directly on untransformed experimental data;
(ii) points can be weighted according to their theoretical
(expected) or experimental (measured) variability;
(iii) the reliability of parameters can be assessed from
their standard deviations which, although approximated, are
statistically more correct than those from the linear
regression analysis on transformed data.
    In order to compare different models fitted to a
given set of experimental points, some criteria for
"goodness of fit" are the value of (weighted) sum of squared errors (SS) and the root mean squared error (rms error
$= \sqrt{SS/(N \text{ points} - N \text{ parameters})}$). If a model with additional parameters (e.g. a two site model) fits data better
(smaller SS) than a simpler one (e.g. a one site model),

the significance of improvement of goodness of fit can be
tested with a F test on the two SS (8);

$$ F = \frac{SS_1 - SS_2}{df_1 - df_2} * \frac{df_2}{SS_2} $$

(where: 1 stands for the simple and 2 for the complex
model; df are degrees of freedom = Number of points -
number of parameters). The calculated F value can be
compared with F table at the desired level of probability,
with $(df_1 - df_2)$ and df_2 degrees of freedom. Also in this
case, the use of Plot od Standardized residuals and Runs
test (see before) is suggested.

ACKNOWLEDGEMENTS

     This study was partially supported by CNR (F.P. Fine
Chemistry) Rome. F.B. is recipient of a grant "Anna Villa
Rusconi" (Varese, Italy). Mrs Franca Barbieri is grate-
fully acknowledged for the excellent secretarial assistance.

REFERENCES

1. J.P. Bennett,Methods in binding studies, in: "Neuro-
     transmitter Receptor Binding", H.I. Yamamura, S.J.
     Enna and M.J. Kuhar, eds., Raven Press, New York
     (1978).
2. G.A. Weiland and P.B. Molinoff, Quantitative analysis
     of drug-receptor interactions. I. Determination of
     kinetic and equilibrium properties, Life Sci. 29:
     313 (1981).
3. S.J. List and P. Seeman, Resolution of dopamine and
     serotonin receptor components of 3H-spiperone binding
     to rat brain regions, Proc. Natl. Acad. Sci. USA 78:
     2620 (1981).
4. Y. Cheng and W.H. Prusoff, Relationship between the
     inhibition constant (Ki) and the concentration of
     inhibitor which causes 50 per cent inhibition (IC50)
     of an enzymatic reaction, Biochem. Pharmacol. 22:
     3099 (1973).
5. J.A. Zivin and D.R. Waud D.R., How to analyze binding,
     enzyme and uptake data: the simplest case, a single
     phase, Life Sci. 30: 1407 (1982).
6. V. Guardabasso, Some mathematical and statistical iss-
     ues in the quantitative analysis of data from binding

site studies. EMBO Workshop "Membrane mechanisms in synaptic transmission", Ponte di Legno (Brescia, Italy), July 1982.

7. T.W. Plumbridge, L.J. Aarons and J.R. Brown, Problems associated with analysis and interpretation of small molecule/macromolecule binding data, J. Pharm. Pharmacol. 30: 69 (1978).

8. P.J. Munson and D. Rodbard, LIGAND: a versatile computerized approach for characterization of ligand-binding systems, Anal. Biochem. 107: 220 (1980).

9. A.K. Thakur, M.L. Jaffe and D. Roodbard, Graphical analysis of Ligand-Binding Systems; evaluation by Montecarlo Studies, Anal. Biochem. 107: 279 (1980)

10. P.B. Molinoff, B.B. Wolfe and G.A. Weiland, Quantitative analysis of drug-receptor interactions. II. Determination of the properties of receptor subtypes, Life Sci. 29: 427 (1981).

11. J.G. Norby, P. Ottolenghi and J. Jensen, Scatchard Plot: common misinterpretation of binding experiments, Anal. Biochem. 102: 318 (1980).

12. R.J. Tallarida and R.B. Murray, "Manual of Pharmacological Calculations", Springer-Verlag, New York (1981)

13. P.K. Minneman, L.R. Hegstrand and P.B. Molinoff, Simultaneous determination of ß-1 and ß-2 adrenergic receptors in tissues containing both receptor subtypes, Mol. Pharmacol. 16: 34 (1979).

14. G. Sacchi-Landriani, V. Guardabasso and M. Rocchetti, NL-FIT: a microcomputer program for non-linear fitting, Computer Progr. in Biomed., 16:35 (1983).

15. H.A. Feldman, Mathematical theory of complex ligand-binding systems at equilibrium: some methods for parameter fitting, Anal. Biochem. 48: 317 (1972).

16. A. De Lean, P.J. Munson and D. Rodbard, Simultaneous analysis of families of sigmoidal curves: application to bioassay, radioligand assay, and physiological dose-response curves, Am. J. Physiol. 235: E97 (1978).

17. G.W. Snedecor and W.G. Cochran, "Statistical Methods", The Iowa State University Press, Ames (USA) (1967).

18. C.A. Bennet and N.L. Franklin, "Statistical Analysis in Chemistry and the Chemical Industry", Wiley, New York (1954).

19. Y. Bard, "Nonlinear Parameter Estimation", Academic Press, New York (1974).

20. C.M. Meltzer, G.L. Elfring and A.J. McEwan, "A Users Manual for NONLIN", The Upjohn Company, Michigan (1974).
21. C.C. Peck and B.B. Barret, Nonlinear least-squares regression programs for minicomputers, J. Pharmacokinetics Biopharm. 7: 537 (1979).
22. W.R. Greco, R.L. Priore, M. Sharma and W. Korytnyk, ROSFIT: an enzyme kinetics nonlinear regression curve fitting package for a microcomputer, Computers and Biomed. Res. 15: 39 (1982).

THE MUSCARINIC RECEPTORS

Giancarlo Pepeu and Ileana Marconcini Pepeu

Department of Pharmacology
University of Florence
Viale Morgagni 65,   50134 Florence, Italy

Muscarinic receptors exist in both peripheral tissues and the central nervous system. Their activation is responsible for all the effects elicited by parasympathetic stimulation in the peripheral organs and modulate ganglionic transmission (Wallis, 1979). In the central nervous system muscarinic receptors are involved in many functions connected with the cognitive processes, sleep mechanism, motility, blood pressure and water intake regulation (Karczmar, 1981).

By measuring the responses of simple organ preparations - usually isolated guinea-pig or rabbit intestine - the structural requirements of the muscarinic agonists and antagonists and their affinity for the receptors have been studied for many years, as reported in several reviews (Waser, 1861; Brimblecombe, 1974).

Presynaptic receptors

Using pharmacological procedures, muscarinic receptors have been identified at pre- and postsynaptic levels. When the muscarinic presynaptic receptors are located on cholinergic nerve endings, they are called "autoreceptors". The presence of muscarinic presynaptic receptors has been described also on monoaminergic fibres (Fozard and Muscholl, 1972). The function of the muscarinic autoreceptors is to modulate the evoked ACh release from peripheral (Kilbinger and Wessler,1980; Alberts et al., 1982) and central neurons (Polak and Meews,1966; Szerb and Somogyi, 1973; Nordström and Bartfai, 1980) The activation of the presynaptic receptors

decreases, while their blockade enhances ACh release from
the nerve endings. Until now the comparison of the effect
of different antagonists with varying potencies has not
indicated any pharmacological variations between central
muscarinic autoreceptors on the one hand and, on the
other, both central and peripheral postsynaptic muscarinic receptors (Bowen and Marek, 1982). By using several
muscarinic agonists, however, Kilbinger and Wessler (1980)
have found slight differences in the pharmacological
properties of the pre- and postsynaptic muscarinic receptors of the guinea-pig ileum.

Ontogenesis

The ontogenesis of muscarinic receptors is rather
slow. In the rat cerebral cortex, muscarinic receptor
concentration as measured by radioligand binding takes
10 days to reach 30% of adult value (Kuhar, et al.,
1980). Scopolamine does not enhance locomotor activity
in rats until after 15 days of life (Campbell, et al.,
1969) and cortical presynaptic autoreceptors are not
functioning in 7-day-old rats (Pedata, et al., 1983).

Receptor mechanisms

A receptor is formed by a recognition site, and
biochemical or biophysical machinery which transduces
the signal elicited by the agonist binding to the recognition site into a physiological effect. The physiological events triggered by the activation of the muscarinic receptors have recently been reviewed by Hartzell (1982). The receptors can be directly coupled to
ionic channels modulating ion fluxes, or they can exert
metabolic effects which, in turn, influence Ca^{2+} fluxes.
A Ca-dependent increase in cGMP levels caused by muscarinic stimulation has been observed in several peripheral
tissues and tissue slices from sympathetic ganglia and
brain (see ref.in Heilbronn and Bartfai, 1978), and in
neuroblastoma cells (Matsuzawa and Nirenberg, 1975).
The increase in cGMP levels activates in turn cyclic
GMP-dependent protein kinases (Kuo, 1974).
A muscarinic receptor-mediated GMP-dependent hyperpolarization of the membrane potential in mouse neuroblastoma cells has been also demonstrated (Wastek, et
al., 1981).
Muscarinic receptors are also involved in phosphatidylinositol breakdown, which is possibly connected to
Ca mobilization (see Mitchell and Kirk, 1982).

Receptor binding studies

Direct measurement of the number of muscarinic receptors and of ligand affinity using receptor binding techniques began in 1974 (Yamamura and Snyder, 1974; Beld and Ariens, 1974) following the pioneer work of Paton and Rang (1965). Many antagonists and agonists have been used as specific ligands. They have been listed by Heilbronn and Barfai (1978). Generally speaking, antagonists show a much higher affinity than agonists for the binding sites. The difference is exemplified by the antagonist quinuclidinyl benzylate (QNB), whose IC_{50} is $4 - 5 \times 10^{-10}$ M, and by the agonist oxotremorine which has an IC_{50} of $5 - 8 \times 10^{-7}$ M (Snyder et al.1975). The binding studies have shown the strict stereospecificity of the muscarinic receptors (Aronstam et al., 1979; Laduron et al., 1979). They make it possible to calculate the number of receptors (B_{max}) as well as their affinity constant (K_m), and their tissue distribution (Yamamura et al., 1974); Wastek and Yamamura, 1978). They also make it possible to visualize the receptors by autoradiographic techniques (Kuhar and Yamamura, 1975).

Equivalence between muscarinic receptors identified by binding studies and physiological techniques

In the case of muscarinic antagonists a linear relationship has been obtained by plotting the affinity constants (or their log) derived from inhibition of the guinea-pig ileum contraction, against the affinity constants (or their log) obtained from binding studies (Birdsall et al., 1977; Hulme et al., 1978; Aronstam et al., 1979).

On the other hand a linear relationship was found in the case of muscarinic agonists by plotting the log of the affinity constant for high affinity binding sites against the reciprocal of agonist concentration which produces a half-maximal physiological response (Birdsall et al., 1977).

The agreement that exists between the affinity constants derived from binding and physiological studies even when the latter are carried out on different animal species shows that both procedures measure the same receptors. On the basis of these studies,Hulme et al. (1978) claim that the ligand-binding subunits of central and peripheral muscarinic receptors are similar, if not identical, and that the functional properties of the receptor have been strongly conserved during the course

of vertebrate evolution. However a discrepancy between
the receptor binding of agonists and their biological
effects has been observed in some instances (Furchgott,
1978) and it has been explained by the existence of
"spare receptors" (Putney and Van der Walle, 1980).

High and low affinity binding sites

The equilibrium binding of muscarinic antagonists
to brain membrane fractions conforms closely to the simplest model, namely that of competitive interaction
with a single uniform set of sites (Hulme et al., 1978).
This model has been supported particularly by the finding that binding capacities for different antagonists
are equal.
When the binding of muscarinic agonists is investigated - either by studying the displacement of a tritiated antagonist from the receptor by increasing the
concentration of an agonist, or by studying the direct
binding of a tritiated agonist (Birdsall and Hulme,
1976) - a different picture emerges. The analysis of
the nonlinear occupancy concentration curves for carbachol demonstrates the presence of two major populations of agonist binding sites with high and low affinity constants. These sites do not interconvert during
the binding experiments, and have the same affinity
constant for antagonists (Birdsall et al., 1978). As
an example the affinity constant of carbachol for the
high affinity sites is 1.7×10^6 M, while for the low
affinity sites is 1.2×10^4 M. The ratio between high
and low affinity constants for different agonists can
vary between 1 and 275, while the percentage of high
affinity sites ranges from 19 for oxotremorine to 49
for propionylcholine. The analysis of the direct binding curve of tritiated oxotremorine M also indicates
the existence of a small number of super high affinity
binding sites with an affinity constant of 3.5×10^8 M
(Birdsall et al., 1978).
The two forms of muscarinic receptors, with low
and high affinity constants, are interconvertible.
Guanine nucleotides can modify the interaction of agonists with muscarinic receptors by lowering the receptor-ligand affinity (Rosenberger et al.,1979;Sokolowsky
et al., 1980). More recently it has been shown in the
frog heart that guanine nucleotides also influence
antagonist binding at muscarinic receptors by increasing the antagonist affinity (Burgisser et al., 1982).
The physiological meaning of this interaction has
not been yet fully understood. Birdsall and Hulme

(1976) speculate that the low affinity site is coupled
to a transducer, whereas the high affinity state represents an uncoupled form of the receptor. The binding
of guanyl nucleotides with a protein connected with the
receptor would therefore initiate the receptor- transducer coupling in the muscarinic, as in other, receptor
systems (Creese et al., 1979).

The interconversion between high and low affinity
muscarinic binding sites can also be triggered by
pharmacological treatments. In rat heart, isoproterenol
decreases the percentage of high affinity muscarinic
receptors. This effect is specifically blocked by propranolol, a fact which suggests an interdependent link-
age between β adrenergic and muscarinic receptors in
cardiac membranes (Rosemberger et al., 1980).

The affinity of the muscarinic receptors in the
cat submandibular salivary gland is markedly increased
by the presence of the vasoactive intestinal peptide
(VIP) (Lundberg et al.,1982). It has been shown
(Lundberg, 1981) that VIP is co-released with ACh from
the secretory nerves and potentiates ACh-induced salivary secretion.

Muscarinic receptor subtypes M_1 and M_2

Pharmacological experiments have shown that
pirenzepine - a new tricyclic compound with anticholinergic properties - causes a 50% inhibition of gastric
acid secretion at concentrations which have little
effect on pupillary diameter (Parry and Heathcote,1982).
In contrast, the amount of atropine required to inhibit
acid secretion by 50% will significantly increase pupil
diameter.

These results support a growing body of evidence
suggesting that pirenzepine is able to distinguish
between different subclasses of muscarinic binding
sites (Hammer et al., 1980). The subclasses have been
defined M_1 - which includes excitatory ganglionic receptors -and M_2 - present in effectory organs like
smooth muscle and heart (Hammer and Giachetti, 1982).

It has also been shown that pirenzepine selectively identifies a high affinity population of muscarinic
receptors in the rat cerebral cortex. Conversely, few
pirenzepine binding sites were demonstrated in the
heart and cerebellum (Watson et al., 1982).

Contrary to these findings, after estimating the
pA_2 for isolated ileum and atrium of guinea-pig and for
acid secretion in mouse, Szeleny (1982) concludes that
pirenzepine is an anticholinergic agent without specific

activity for gastric muscarinic receptors. Therefore the
concept of distinct multiple muscarinic receptors must
still be considered with some caution.

Muscarinic receptor regulation

The number and affinity of muscarinic receptors
is regulated by their ligand and by drug treatment. In
the rat brain, chronic treatment with cholinesterase
inhibitors brings about a decrease in the number of
binding sites, ranging from 10% in the hypothalamus to
50% in the striatum (Gazit et al., 1979; Costa et al.,
1982). According to Ehlert et al.(1980), following this
treatment the muscarinic receptors display a greater
decrease in affinity to agonists than to antagonists.
The changes in receptor number and affinity can be considered the biochemical basis of the tolerance which
develops during chronic anticholinesterase drug administration.
Conversely, the repeated administration of anticholinergic drugs is followed by an increase in the
number of binding sites, ranging from 21% in the hippo-
campus of young rats (Ben-Barak and Dudai, 1980) to
approximately 40% in the cortex and striatum of adult
rats (Majocha and Baldessarini, 1980) treated for 20
days or 2 weeks, respectively, with large doses of
scopolamine. According to McKinney and Coyle (1982),
the increase in binding sites involves mostly the high
affinity sites.
The down regulation of the muscarinic receptors
can also be induced in cultured neuroblastoma and
embrionic cerebrum cells, in response to receptor
saturating concentration of cholinomimetic drugs. Up
regulation can be induced by atropine only in central
nervous system cultures, and probably depends upon the
blockade of endogenous regulation (Simon and Klein,
1979). The down regulation of the binding sites also
affects the biochemical response to receptor activation;
this is shown by a decrease in the capacity of carbachol
to stimulate phosphatidylinositol turnover, a receptor
mediated response (Simon and Klein, 1981).
A significant increase in muscarinic binding sites
has been found in the striatum and brain stem of rats
chronically treated with barbiturates, 3 days after
withdrawal (Nordberg et al., 1980).
An increased affinity in muscarinic binding sites
was found (Casamenti et al., 1980) in the cortex and
striatum of rats made tolerant to morphine. The increase
disappeared during naloxone-induced withdrawal syndrome,

and was also associated with a persistent decrease in
the number of binding sites in the striatum only.

A significant decrease in muscarinic binding sites
in the cerebellum of 21-day-old hyperthyroid rats, and
an increase in thyroid deficient rats (Patel et al.,
1980) was also noted.

All these results taken together demonstrate that
receptor plasticity represents an important adaptive
mechanism involved in a variety of conditions affecting
central and peripheral cholinergic transmission. Up and
down regulation of muscarinic receptors can be considered a long-term postsynaptic modulatory mechanism of
the ACh-mediated impulse flow.

Changes in muscarinic binding sites during diseases and aging

In human brain a substantial decrease in muscarinic
binding sites, associated with a decrease in choline
acetyltransferase activity, has been found in the
striatum of patients affected by Huntington's chorea
(Wastek and Yamamura, 1978).

A decrease in muscarinic receptors is a common
finding in the brain of aged rodents. However in brains
of elderly humans controversial findings are reported,
particularly in patients affected by senile dementia of
Alzheimer type. According to the extensive review of
the literature made by Bartus et al.(1982) in the majority of the investigations, no decrease in receptor
density has been found. It should be mentioned, however,
that in rats a lesion of the nucleus basalis magno-
cellularis, which removes a large part of the cholinergic input to the cerebral cortex, induces a small
decrease of the binding sites, followed within 3 weeks
by a return to normal level (McKinney and Coyle,1982).
It is no surprise, therefore, that in human aging brain,
where the changes in cholinergic functions occur over
a period of years, modifications in the number of muscarinic binding sites are so difficult to detect.

Muscarinic receptor axonal transport

The measurement and visualization of muscarinic
binding sites has made it possible to detect an accumulation of muscarinic receptors on both sides of a lig-
ature placed around dog splenic nerve (Laduron,1980),
and both rat sciatic (Wamsley et al.,1981) and vagus
(Zarbin et al.,1982) nerves. According to the latter
authors, most of the binding sites which accumulate

proximal to ligatures bind to the agonist carbachol
with high affinity, while most of the sites accumulating
distally bind carbachol with a low affinity.

A cycle may be envisaged involving the synthesis
of muscarinic receptors in the perikaria, an antero-
grade flow in the axons with externalization at terminals
by exocytosis. The receptors become functional, and are
then internalized and transported by retrograde axonal
flow to the perikaria, where they are recycled or
degraded. The precise physiological role of receptor
axonal transport and the meaning of the affinity changes,
however, have yet to be established.

Conclusion

In concluding this short overview of the vast
literature on the many facets of the muscarinic receptors,
I would like to point out that our knowledge is far from
exhaustive, particularly with regard to the biochemical
and biophysical events triggered by muscarinic receptor
activation. Furthermore the differences between the
various types of muscarinic receptors may be larger
than we now think. Nördström et al.(1983) have recently
described a muscarinic cholinergic ligand which acts as
both a presynaptic antagonist and a postsynaptic agonist.
The synthesis of new therapeutic agents can therefore
be envisaged.

REFERENCES

Alberts, P., Bartfai, T. and Stjärne, L., 1982, The
effects of atropine on ^{3}H acetylcholine secretion from
guinea-pig myenteric plexus evoked electrically or by
high potassium, J. Physiol.(Lond.), 329: 93.

Aronstam, R.S., Triggle,D.J. and Eldefrawi, M.E., 1979,
Structural and stereochemical requirements for muscarinic receptor binding, Mol. Pharmacol., 15 : 227.

Bartus, R.T., Dean, R.L., Beer, B. and Lippa, A.S.,
1982, The cholinergic hypothesis of geriatric memory
disfunction, Science, 217 : 408.

Beld, A.J. and Ariens, E.J., 1974, Stereospecific
binding as a tool in attempts to localize and isolate
muscarinic receptors, Eur. J. Pharmac., 25 : 203.

Ben-Barak, J. and Dudai, Y., 1980, Scopolamine induces
an increase in muscarinic receptor level in rat hippo-
campus, Brain Res., 193 : 309.

Birdsall, N.J.M.,and Hulme, E.C., 1976, Biochemical studies on muscarinic acetylcholine receptors, J. Neurochem., 27 : 7.

Birdsall, N.J.M., Burgen, A.S.V.,and Hulme, E.C., 1977, Correlation between the binding properties and pharmacological responses of muscarinic receptors, in : "Cholinergic mechanisms and psychopharmacology", D.J. Jenden, ed., Plenum Press, New York.

Birdsall, N.J.M., Burgen, A.S.V.,and Hulme, E.C., 1978, The binding of agonists to brain muscarinic receptors, Mol. Pharmacol., 14 : 723.

Bowen, D.M. and Marek, K.L., 1982, Evidence for the pharmacological similarity between central presynaptic muscarinic autoreceptors and postsynaptic muscarinic receptors, Br. J. Pharmac., 75 : 367.

Brimblecombe, R.W., 1974,"Drug action on cholinergic systems", The Macmillan Press, London.

Burgisser, E., De Lean, A.,and Lefkowitz, R.J., 1982, Reciprocal modulation of agonist and antagonist binding to muscarinic cholinergic receptor by guanine nucleotide, Proc. Natl. Acad. Sci. U.S.A., 79 : 1732.

Campbell, B.A., Lytle, L.D., and Fibiger, H.C., 1969, Ontogeny of adrenergic arousal and cholinergic inhibitory mechanisms in the rat, Science, 166 : 635

Casamenti, F., Pedata,F., Corradetti, R., and Pepeu, G., 1980, Acetylcholine output from the cerebral cortex, choline uptake and muscarinic receptors in morphine-dependent freely-moving rats, Neuropharmacol.,19 : 597.

Costa, L.G., Schwab, B.W., and Murphy, S.D., 1982, Differential alterations of cholinergic muscarinic receptors during chronic and acute tolerance to organophosphorus insecticides, Biochem. Pharmacol., 31 : 3407.

Creese, I., Usdin, T.B., and Snyder, S.H., 1979, Dopamine receptor binding regulated by guanine nucleotides, Mol. Pharmacol., 16 : 69.

Ehlert, F.J., Kokka, N., and Fairhurst, A.S., 1980, Altered ^{3}H quinuclidinyl benzylate binding in the striatum following chronic cholinesterase inhibition with diisopropylfluorophosphate, Mol. Pharmacol., 17 : 24.

Fozard, J.R. and Muscholl, E., 1972, Effects of several muscarinic agonists on cardiac performance and the release of noradrenaline from sympathetic nerves of the perfused rabbit heart, Br. J. Pharmac., 45 : 616.

Furchgott, R.F., 1978, Pharmacological characterization of receptors: its relation to radioligand-binding studies, Fedn. Proc., 37 : 115.

Gazit, H., Silman, I., and Dudai, Y., 1979, Administration of an organophosphate causes a decrease in mus - carinic receptor level in rat brain, Brain Res., 174 : 351.

Hammer, R., Berrie, C.P., Birdsall, N.J.M., Burgen, A. S.V., and Hulme, E.C., 1980, Pirenzepine distinguishes between different subclasses of muscarinic receptors, Nature, 283 : 90.

Hammer, R. and Giachetti, A., 1982, Muscarinic receptor subtypes: M_1 and M_2. Biochemical and functional characterization, Life Sci., 31 : 2991.

Hartzell, H.C., 1982, Physiological consequences of muscarinic receptor activation, Trends Pharmacol. Sci., 3 : 213.

Heilbronn, E. and Bartfai, T., 1978, Muscarinic acetylcholine receptors, Progr. Neurobiol., 11 : 171.

Hulme, E.C., Birdsall, N.J.M., Burgen, A.S.V., and Mehta, P., 1978, The binding of antagonists to brain muscarinic receptors, Mol. Pharmacol., 14 : 737.

Karczmar, A.G., 1981, Basic phenomena underlying novel use of cholinergic agents, anticholinesterases and precursors in neurological including peripheral and psychiatric diseases, in : "Cholinergic mechanisms", G. Pepeu and H. Ladinsky eds., Plenum Press, New York.

Kilbinger, H. and Wessler, I., 1980, Pre and postsynaptic effects of muscarinic agonists in the guineapig ileaum, Naunyn-Schmiedeberg's Arch. Pharmacol., 314 : 259.

Kuhar, M.J. and Yamamura, H.I., 1975, Light autoradiographic localization of cholinergic muscarinic receptors in rat brain by specific binding of a potent antagonist, Nature, 253 : 560.

Kuhar, M.J., Birdsall, N.J.M., Burgen, A.S.V. and Hulme, E.C., 1980, Ontogeny of muscarinic receptors in rat brain, Brain Res., 184 : 375.

Kuo, J.F., 1974, Guanosine 3' 5'- monophosphate-dependent protein kinases in mammalian tissues, Proc. Natl. Acad. Sci. U.S.A., 71 : 4037.

Laduron, P., 1980, Axoplasmic transport of muscarinic receptors, Nature, 286 : 287.

Laduron, P., Verwimp, M, and Leysen, J.E., 1979,

Stereospecific in vitro binding of ^{3}H dexetimide to brain muscarinic receptors, J. Neurochem., 32 : 421.

Lundberg, J.M., 1981, Evidence for coesistence of vasoactive intestinal polypeptide (VIP) and acetylcholine in neurons of cat exocrine glands, Acta Physiol. Scand., Suppl. 496 : 1.

Lundberg, J.M., Hedlund, B., and Bartfai, T., 1982, Vasoactive intestinal polypeptide enhances muscarinic ligand binding in cat submandibular salivary gland, Nature, 295 : 147.

Majocha, R. and Baldessarini, R.J., 1980, Increased muscarinic receptor binding in rat forebrain after scopolamine, European J. Pharmacol., 67 : 327.

Matsuzawa, H. and Nirenberg, M., 1975, Receptor mediated shift in cGMP and cAMP levels in neuroblastoma cells, Proc. Natl. Acad. Sci. U.S.A., 72 : 3472.

McKinney and Coyle, J.T., 1982, Regulation of neocortical muscarinic receptors: effect of drug treatment and lesions, J. Neurosci., 2 : 97.

Mitchell, R.H. and Kirk, C.J., 1981, Why is phosphatidylinositol degraded in responses to stimulation of certain receptors ?, Trends Pharmacol. Sci., 2 : 86.

Nordberg, A., Wahlström, G and Larsson, Ch., 1980, Increased number of muscarinic binding sites in brain following chronic barbiturate treatment to rat, Life Sci., 26 : 231.

Nordström, O. and Bartfai, T., 1980, Muscarinic autoreceptor regulates acetylcholine in rat hippocampus: in vitro evidence, Acta Physiol., 108 : 347.

Nordström, O;, Alberts, P., Westlind, A., Unden, A., and Bartfai, T., 1983, Presynaptic antagonist – postsynaptic agonist at muscarinic cholinergic synapses: N-methyl-N-(1-methyl-4-pyrrolidino-2-butynyl) acetamide, Mol. Pharmacol., in the press.

Parry, M. and Heathcote, B.V., 1982, A comparison of the effec of pirenzepine and atropine on gastric acid secretion, salivary secretion and pupil diameter in the rat. Life Sci., 31 : 1465.

Patel, A.J., Smith, R.M., Kingsbury, A.E., Hunt, A., and Balazs, R., 1980, Effects of thyroid state on brain development: muscarinic acetylcholine and GABA receptors, Brain Res., 198 : 389.

Paton, W.D.M. and Rang, H.P., 1965, The uptake of atropine and related drugs by intestinal smooth muscle of

the guinea-pig in relation to acetylcholine receptors,
Proc. R. Soc. B, 163 : 1.

Pedata, F., Slavikova, J., Kotas,A., and Pepeu, G., 1983,
Acetylcholine release from rat cortical slices during
postnatal development and aging, Neurobiol. Aging,
in the press.

Polak, R.L. and Meeuws, M.M., 1966, The influence of
atropine on the release and uptake of acetylcholine by
the isolated cerebral cortex of the rat, Biochem.
Pharmacol., 15 : 989.

Putney, J.W. and Van De Walle, C., 1980, The relationship between muscarinic receptor binding and ion move-
ments in rat parotid cells, J. Physiol. (Lond.), 299 :
521.

Rosenberger,L.B.,Roeske, W.R., and Yamamura, H.I., 1979,
The regulation of muscarinic receptors by guanine nucleotides in cardiac tissue, European J. Pharmacol., 56 : 179.

Rosenberger, L.B., Yamamura, H.I., and Roeske, W.R., 1980
The regulation of cardiac muscarinic receptors by isoproterenol, European J. Pharmacol., 65 : 129.

Siman, R.G. and Klein, W.L., 1981, Specificity of muscarinic acetylcholine receptor regulation by receptor
activity, J. Neurochem., 37 : 1099.

Siman, R.G., and Klein, W.L., 1979, Cholinergic activity
regulates muscarinic receptors in central nervous system
cultures, Proc. Natl. Acad. Sci. U.S.A., 76:4141.

Snyder, S.H., Chang, K.J., Kuhar, M.J., and Yamamura,
H.I., 1975, Biochemical identification of the mammalian
muscarinic cholinergic receptor, Fedn. Proc., 34 : 1915.

Sokolovsky, M., Gurwitz, D., and Galron, R., 1980,
Muscarinic receptor binding in mouse brain: regulation
by guanine nucleotides, Biochem. Biophys. Res. Comm.,
94 : 487.

Szelenyi, I., 1982, Does pirenzepine distinguish
between "subtypes" of muscarinic receptors ?, Br. J.
Pharmac., 77 : 567.

Szerb, J.C. and Somogyi, C.T., 1973, Depression of
acetylcholine release in cerebral cortical slices by
cholinesterase inhibition and by oxotremorine, Nature,
241 : 121.

Wallis, D.I., 1979, The apparent multiplicity of ganglionic receptors, in:"Trends in autonomic pharmacology",
I, Kalsner ed., Urban and Schwarzenberg, Baltimore.

Wamsley, J.K., Zarbin, M.A., and Kuhar, M.J., 1981,

Muscarinic cholinergic receptors flow in the sciatic nerve, Brain Res., 217 : 155.

Waser, P.G., 1961, Chemistry and pharmacology of muscarine, muscarone and related compounds, Pharmacol. Rev., 13 : 465.

Wastek, G.J. and Yamamura, H.I., 1978, Biochemical characterization of the muscarinic cholinergic receptor in human brain: alteration in Huntington's disease, Mol. Pharmacol., 14 : 768.

Wastek, G.J., Lopez, J.R., and Richelson, E., 1981, Demonstration of a muscarinic mediated cyclic GMP-dependent hyperpolarization of the membrane potential of mouse neuroblastoma cells using ^{3}H tetraphenylphosphonium, Mol. Pharmacol., 19 : 15.

Watson, M., Roeske, W.R., and Yamamura, H.I., 1982, ^{3}H pirenzepine selectively identifies a high affinity population of muscarinic cholinergic receptors in the rat cerebral cortex, Life Sci. 31 : 2019.

Yamamura, H.I. and Snyder, S.H., 1974, Muscarinic cholinergic binding in rat brain, Proc. Natl. Acad. Sci. U.S.A., 71 : 1725.

Yamamura, H.I., Kuhar, M.J., Greenberg, D., and Snyder, S.H., 1974, Muscarinic cholinergic receptor binding: regional distribution in monkey brain, Brain Res., 66: 541.

Zarbin, M.A., Wamsley, J.K., and Kuhar, M.J., 1982, Axonal transport of muscarinic cholinergic receptors in rat vagus nerve: high and low affinity agonist receptors move in opposite directions and differ in nucleotide sensitivity, J. Neurosci., 2 : 934.

THE β-ADRENERGIC RECEPTORS AND THEIR MODE

OF INTERACTION WITH ADENYLATE CYCLASE

Alexander Levitzki

Department of Biological Chemistry
The Hebrew University of Jerusalem
91904 Jerusalem, Israel

1. *Classification of catecholamine receptors*

The classification into α- and β-receptors has been extended in recent years as a result of extensive detailed studies of structure-activity relationships; thus, β-adrenergic responses can be divided into two subclasses: β_1 and β_2. Epinephrine and norepinephrine are approximately equipotent agonists for β_1-receptors which are preferentially blocked by practolol. Epinephrine is more potent than norepinephrine in activating β_2-receptors which are inhibited by butoxamine. A variety of antagonists such as propranolol, alprenolol and pindolol inhibit the catecholamine response of both β_1- and β_2-receptors. β_1 and β_2 are functionally similar as both of them are coupled to and activate the enzyme adenylate cyclase as the signal system, and will therefore be discussed as a single family of β-receptors.

α-Adrenergic receptors are different, ont only in their ligand specificity but also functionally. α_1-Receptors are the characteristic postsynaptic α-adrenergic receptors, whereas α_2-receptors are either presynaptic receptors in catecholaminergic synapses or postsynaptic receptors in the periphery. Presynaptic α_2-receptors regulate the release of neurotransmitters, including norepinephrine. α_1-Adrenergic receptors are coupled to specific Ca^{2+} channels and α_2-adrenergic receptors inhibit adenylate cyclase.

Both α- and β-adrenergic receptors are stereospecific for the R stereoisomer for both the agonist and the antagonist.

A third major class of catecholamine receptors are the dopamine receptors which occur in the central nervous system and in the peri-

79

phery, and in this case the agonist specificity is dopamine >>
norepinephrine. The dopaminergic response is blocked completely
by specific antagonists such as phenothiazines, butyrophenones and
butaclamol. There are two confirmed classes of depamine receptors,
D1-receptors and D2-receptors, which differ in their ligand speci-
ficities as well as in their biochemical properties. The D1-recep-
tor is coupled to adenylate cyclase, whereas the D2-receptor is not.

2. *The activation of adenylate cyclase by β-receptors*

Of all catecholamine receptors the β-adrenergic receptors have
received the most attention. This is due to the fact that the primary biochemical signal elicited upon agonist binding is the activation of adenylate cyclase to produce the "second messenger" cyclic
AMP (cAMP) from ATP with the target cell:

$$\text{ATP} \xrightarrow[\text{Mg}^{2+},\ \text{GTP}]{\text{catecholamine}} \text{cyclic AMP} + \text{PPi} \qquad (1)$$

The coupling of β-adrenergic receptors with adenylate cyclase
is similar to the coupling between adenylate cyclase and the receptors for polypeptide hormones such as glucagon, ACTH and secretin.
In certain cells such as the liver cell and the fat cell, both β-
-adrenergic receptors and receptors for polypeptide hormones are
coupled to the enzyme adenylate cyclase. The activation of adenylate cyclase by β-adrenergic agonists requires the nucleotide GTP
which acts in a synergistic fashion with the catecholamines (for
review, see Ref. 1). The degree of occupancy of the β-adrenergic
receptor with the agonist and occupancy of the GTP regulatory site
by intracellular GTP determine the final output of cAMP by the
adenylate cyclase. The second messenger cAMP produced intracellu-
larly by the enzyme triggers many biochemical events characteristic
to the cell, the activation of protein kinase being the first step.
Other biochemical responses which are cAMP independent and are apparently coupled to the β-receptor, are: catecholamine dependent
Ca^{2+} efflux, inhibition of Mg^{2+} efflux, and the carboxymethylation
of phosphatidyl-ethanolamine. Less is known about the significance
of these biochemical reactions coupled to the β-receptors.

3. *Radioligand binding assay of β-adrenergic receptors*

In 1974 it was firmly established that catecholamine binding
is an inadequate monitor of authentic β-adrenergic receptors for
two reasons: (a) The receptor concentration accessible experimentally (up to 10 nM) is far below the catecholamine-receptor dissociation constant (100 nM to 100 μM). Furthermore, the high dissociation constants are mainly due to a rapid rate of dissociation
of the receptor-ligand complex. Therefore, a filter assay cannot
be used, as the life time of the receptor-ligand complex is much
too short even for the shortest filtration time attainable experimentally. (b) Catecholamines bind to many non-receptor components

in the membrane preparations studied, and the non-specific catecholamine binding constitutes over 90% of the total (1). In 1974,
radioactively labeled β-adrenergic antagonists were found to monitor
β-adrenergic receptors reliably. The first ligand used was [^{3}H]-
propranolol; shortly thereafter and independently, [^{125}I]-labeled
hydroxybenzylpindolol and [^{3}H]dihydroalprenolol were introduced.
More recently, [^{125}I]-labeled (−)-pindolol and [^{125}I]-labeled (−)-
-cyanopindolol were found to be extremely potent and selective markers with dissociation constants of of 50 pM and 30 pM respectively
for β-adrenergic receptors. It is likely that these ligands will
quickly replace [^{125}I]-labeled hydroxybenzylpindolol, since they
exhibit a much lower extent of "non-specific" binding and higher
selectivity towards β-receptors. The use of these radioactively
labeled ligands has since become a routine procedure for monitoring
β-adrenergic receptors in a large variety of cells. The affinity
of these β-blockers for the β-receptors is in the picomolar to the
nanomolar concentration range, being 3 to 6 orders of magnitude
higher than the affinity of β-agonists (Table I). Therefore, with
these compounds one can monitor low concentrations of these receptors.

Table I: The Affinity of Different β-Blockers towards
           the β-Receptors

β-Blocker	K_{diss} (nM)	
	From kinetics*	From binding experiments
(−)-Propranolol	1.3 ± 0.1	1.2 ± 0.1
(−)-Alprenolol	10 ± 1	10 ± 1
(−)-Iodohydroxyben- zylpindolol	−	0.05
(−)-Iodopindolol	0.05	0.05
(−)-Cyanopindolol	0.03	0.03

The values quoted represent the range of constants reported
between 1974 and 1979. The [^{125}I]-label in the [^{125}I]hydroxyben-
zylpindolol is localized on the 3-position of the indole ring, as
was recently demonstrated. Previously it was believed that the
label is *ortho* to the OH group on the phenol side chain.

The binding of these compounds to the β-receptors is stereo-
selective for the R stereoisomer, and therefore the R-antagonists
are displaced from the β-receptor specifically by R-(−)-catechola-
mines. It is not clear, however, whether S-(+)-catecholamines bind

---

* Competitive inhibition of (−)-catecholamines in the adenylate
  cyclase reaction.

to β-receptors with a 100-fold less affinity or the S-(+) catecholamines are actually completely ineffective but contaminated with
R-(−)-catecholamine. The dissociation constants found for these
β-blockers using binding experiments match very closely the inhibition constants found from their competitive inhibition with catecholamines in the assay of the β-receptor dependent adenylate
cyclase reaction.

Radioactively labeled antagonists can, in principle, also be
used to monitor detergent solubilized receptors, provided the receptor does not denature and its binding site is not masked by the
detergent in the process of solubilization. Such experiments have
recently been reported, in which digitonin solubilized β-adrenergic
receptors from frog erythrocyte membranes were monitored with $[^3H]$-
dihydroalprenolol. Since ^{125}I-labeled hydroxybenzylpindolol ex-
hibits extremely high affinity toward β-adrenergic receptors, it
can be used to measure these receptors in intact cells. Recently,
the high affinity agonist $[^3H]$hydroxybenzylisoproterenol was used
successfully to monitor β-adrenergic receptors. The availability
of specific ligands for β-adrenergic receptors has also made it
possible to monitor changes in the receptor due to desensitization.

4. *Affinity labeling of the β-adrenergic receptor*

Affinity labeling of the β-adrenergic receptor has recently
been achieved by using irreversible β-blockers such as N-(n-hydro-
xy-3-napthyloxypropyl)-N'-bromoacetylethylenediamine. It inhibits
irreversibly the epinephrine dependent adenylate cyclase activity
without damaging the F^--dependent activity in turkey erythrocyte
membranes.

Propranolol and (−)-epinephrine offer protection against this
affinity labeling reaction. The loss of epinephrine dependent activity is accompanied by the loss of $[^3H]$propranolol binding and
of $[^{125}I]$iodohydroxybenzylpindolol binding. More recently, this
affinity label was prepared in a tritiated form.   When the membranes of the affinity labeled turkey erythrocytes or of the affinity labeled L6P muscle cells were solubilized and subjected to
SDS gel electrophoresis, two tritium-labeled protein bands could
be identified: 37,000 and 41,000 in molecular weight. From these
experiments, it was concluded that these two bands probably represent subunits of the β-adrenergic receptor. The labeled protein constituted only a small fraction of the total membrane protein and was varely detected by protein stain. It should also be
pointed out that proteins of similar electrophoretic mobility are
labeled by the affinity label in both L6P cells and in turkey erythrocytes. This observation suggests that the structure of β-
-adrenergic receptors is similar in different types of cells and
may have been conserved during evolution. More recent experiments
in our laboratory suggest that the receptor in its native state is

an oligomer of 40,000 molecular weight (Feder and Levitzki, in preparation).

5. *Solubilization and purification of the β-adrenergic receptor*

As mentioned above, attempts are being made to solubilize and
purify the β-adrenergic receptor. However, the absolut amount of
β-receptor obtained from turkey erythrocytes or frog erythrocytes
is extremely small to preserve its litand binding ability. Much
work is obviously required before relatively pure β-receptor is
available in quantities sufficient to perform biochemical studies.
Successful attempts to reconstitute the solubilized and purified
β-receptor into an active adenylate cyclase have not as yet been
reported. Such experiments are apparently feasible, since reconstitution of hormone sensitive adenylate cyclase from solubilized
crude components can now be achieved.

6. *Mode of coupling between β-receptors and adenylate cyclase*

Over the past few years it has become apparent that the receptor and the catalytic moiety are not the only components of the
receptor-cyclase complex. Mainly through the pioneering studies of
Rodbell and his colleagues, it became apparent that a third component, the transducer, plays a decisive role in the processing of
the hormonal signal. It turns out that the binding of hormone or
of neurotransmitter is a necessary event for the activation of
adenylate cyclase but is not sufficient. The nucleotide GTP must
be present for the hormone-induced activation of adenylate cyclase
to take place. The role of the GTP regulatory unit, N(G/F), was
not recognized for a long time, probably because the substrate ATP
used in the cyclase assay is usually contaminated with enough GTP
to saturate the GTP regulatory site that binds the nucleotide with
an affinity constant in the submicromolar range. Indeed, in the
turkey erythrocyte system, the affinity of GTP to its regulatory
site on N(G/F) was found to be 1.4×10^{-8} M, and in the human platelet adenylate cyclase 4.0×10^{-8} M.

It is now generally accepted that GTP functions as an intracellular regulator which interacts with a specific regulatory site
on the receptor-cyclase system and activates the enzyme together
with the hormone in a synergistic manner. It was also found to be
generally true that GTP analogs such as Gpp(NH)p, GT₹γS and GpCH₂pp
activate the hormone dependent adenylate cyclase in a quasi-irreversi-
ble fashion and, in the presence of hormone, induce the formation of
a highly active and extremely stable adenylate cyclase. Detailed
kinetic analysis of the β-adrenergic-dependent adenylate cyclases
reveal that the role of the agonist is to facilitate the activation
of the adenylate cyclase by the guanyl nucleotide. Antagonists such
as l-propranolol have no effect on the rate of cyclase activation.

The activation of adenylate cyclase requires the simultaneous binding of the agonist and of the guanyl nucleotide to their respective sites. When both sites are occupied, the enzyme is converted from its inactive E state to its activated state E'. There is evidence in the turkey erythrocyte β-receptor system and in the PGE_1 receptor system of the human platelet, that termination of the hormonal signal occurs concomitantly with the hydrolysis of GTP at the guanyl nucleotide regulatory site to GDP and inorganic phosphate. The hydrolysis step *per se* is independent of the continued presence of the agonist at the receptor site. The enzyme, however, cannot be reactivated until a new molecule of GTP binds to the guanyl nucleotide regulatory site, a process that can occur only if the receptor is occupied by an agonist and if the GDP produced in the GTPase step is removed from the GTP regulatory site.

The possibility was raised that the rate limiting step in hormone induced cyclase activation is the dissociation of GDP from the GTP regulatory site and that the role of the hormone is to facilitate this step. According to this hypothesis, basal activity reflects the hormone independent rate of GDP dissociation from the GTP regulatory site. It follows that the different basal activities exhibited by different cyclase systems reflect different rates of hormone independent GDP dissociation. Experimental data to support this notion was recently published but was refuted on an alternative analysis of the same data. Data from our laboratory demonstrate that GDP dissociation is not rate-limiting in the overall process of hormone activation.

Figure 1 summarizes the cycle of adenylate cyclase activation by hormone and GTP. Thus the fraction of the total cyclase pool in the active form is determined by the ratio k_{off}/k_{on}. The kinetic constant k_{on} depends on hormone concentration, and its maximal value is attained by saturating agonist concentration. The first order rate constant k_{off}, depicting the decay of the activated state to its inactive form, represents the GTPase step at the regulatory site. For the turkey erythrocyte β-adrenergic dependent adenylate cyclase, the values of k_{on} and k_{off} were measured directly. k_{on} was determined by following the rate of adenylate cyclase activation in the presence of Gpp(NH)p: Under these conditions, $k_{off} = 0$, since Gpp(NH)p cannot be hydrolyzed and therefore *all* the cyclase molecules are converted to the activated form of the enzyme. The activation rate k_{on} was found to be in the range of $0.4 - 1.0$ min^{-1} at $37°C$. The value of k_{off} was also measured directly by two methods: The enzyme is incubated with hormone, GTP and non-radioactive ATP; at zero time, an excess of a β-antagonist or the GTP antagonist GDPβS and α-$[^{32}P]$ATP are added simultaneously. At this time, E cannot be reconverted to E' since this conversion requires the continuous presence of agonist at the receptor site; hence, E' decays to E with the characteristic k_{off}. The time course of cAMP-$[^{32}P]$ formation reflects the time course of the decay of E' to E.

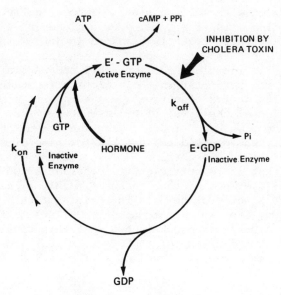

Fig. 1: The on-off cycle of adenylate cyclase activation by
        hormones and GTP.

The direct measurements of k_{on} and k_{off} allow one to determine
whether the model depicted in Figure 1 is satisfactory to describe
the mechanism of adenylate cyclase activation. This model predicts
that the parameters k_{on} and k_{off} will be connected by the relation:

$$V_{max}^{(GTP)} = \frac{V_{max}^{(GppNHp)}}{1 + k_{off}/k_{on}} \tag{2}$$

where $V_{max}^{(GTP)}$ is the maximal specific activity achieved with GTP and
hormone and $V_{max}^{(GppNHp)}$ represents the maximal specific activity attainable with Gpp(NH)p and hormone. Since all the four parameters
depicted in the above equation can be determined independently, it
is possible to *calculate* the value of any of the parameters and to
compare it to the experimentally determined value. In the turkey
erythrocyte system, eq. 2 and therefore the model presented in Fig.
1 were found to correlate well. This technique of measuring k_{on} and
k_{off} values was also successfully used for the PGE$_1$-dependent cyclase
of human platelets.

Partial agonists induce a smaller fraction of the cyclase to be converted into its active form E', thus yielding a lower specific activity as compared with full agonists. Theoretically, this may result either from a lower k_{on} value or a higher k_{off} value or both. These alternatives were recently examined in detail in the case of turkey erythrocyte β-receptor dependent adenylate cyclase. It was found that k_{off} is identical for nine full and partial agonists, whereas k_{on} is the parameter which is agonist dependent. As it was also found in the latter case that $k_{off} \gg k_{on}$, it became apparent from equation 2 that the level of cyclase activation, $v_{max}^{(GTP)}$ is linearly dependent on k_{on}. Indeed, when the steady-state level of cyclase activity was plotted against k_{on} for nine different agonists, a linear relationship was obtained. This finding corroborates further the view that it is sufficient to consider a two-state model for the cyclase system. Another recent finding is that ADP-ribosylation of the GTP binding protein is the origin of the increase in the activity of adenylate cyclase produced by cholera toxin (E' according to our notation). It can be shown that this covalent modification results in the decrease of the GTPase (k_{off}) and thus yields an increase in the steady-state level of active cyclase in the presence of GTP.

7. *The catalytic role of the receptor*

Although it is quite well established that the three basic components of the adenylate cyclase system: the receptor R, the GTP binding protein N(G/F), and the catalytic unit C represent separate macromolecules, little is known about their organization within the membrane, their stoichiometry and the mode of coupling between them. Certain theoretical arguments favor the assumption that the stoichiometry is close to 1:1:1, but no direct data are as yet available. However, more has been learned recently about the mechanism of coupling between these components in one experimental system, the turkey erythrocyte β-adrenergic system. The β-receptor usually separates rather easily from the cyclase. This separation of the cyclase from the β-receptor upon membrane solubilization has been documented in a number of experimental systems. In each case, the solubilized cyclase could respond to NaF, indicating that the catalytic moiety and the GTP regulatory protein remain associated with an as yet unknown stoichiometry. Since the structural work on hormone activated cyclases is progressing slowly, the approach taken to study the mode of coupling between receptors and adenylate cyclase has been mainly a kinetic one. We shall first consider the various theoretical models which have been formulated and then the experiments designed to explore the validity of these models.

In the presence of the non-hydrolyzable analogs of GTP, the N(G/F) and C components are strongly associated and appear as a complex. This complex has recently been partially purified and was found to be of 220,000 molecular weight. In the absence of GTP or its analogs, N(G/F) associates with R. The latter finding was taken to indicate that the N(G/F) unit is activated by the receptor and then, when activated, interacts with C to activate the latter, where N(G/F) physically shuttles from R to C. The difficulties with this model will be discussed below.

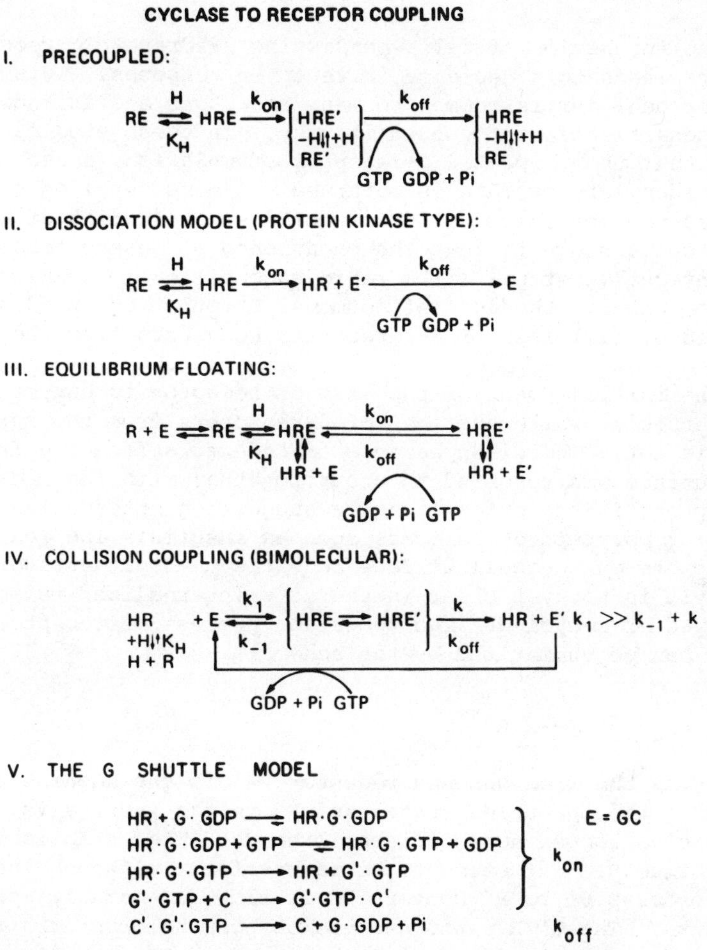

CYCLASE TO RECEPTOR COUPLING

I. PRECOUPLED:

$$RE \underset{K_H}{\overset{H}{\rightleftharpoons}} HRE \xrightarrow{k_{on}} \left\{ \begin{array}{c} HRE' \\ -H\updownarrow+H \\ RE' \end{array} \right\} \xrightarrow{k_{off}} \left\{ \begin{array}{c} HRE \\ -H\updownarrow+H \\ RE \end{array} \right.$$
$$GTP \quad GDP + Pi$$

II. DISSOCIATION MODEL (PROTEIN KINASE TYPE):

$$RE \underset{K_H}{\overset{H}{\rightleftharpoons}} HRE \xrightarrow{k_{on}} HR + E' \xrightarrow{k_{off}} E$$
$$GTP \quad GDP + Pi$$

III. EQUILIBRIUM FLOATING:

$$R + E \rightleftharpoons RE \underset{K_H}{\overset{H}{\rightleftharpoons}} HRE \underset{k_{off}}{\overset{k_{on}}{\rightleftharpoons}} HRE'$$
$$\updownarrow \qquad\qquad \updownarrow$$
$$HR + E \qquad\qquad HR + E'$$
$$GDP + Pi \quad GTP$$

IV. COLLISION COUPLING (BIMOLECULAR):

$$\begin{array}{c} HR \\ +H\updownarrow K_H \\ H + R \end{array} + E \underset{k_{-1}}{\overset{k_1}{\rightleftharpoons}} \left\{ HRE \rightleftharpoons HRE' \right\} \overset{k}{\underset{k_{off}}{\longrightarrow}} HR + E' \quad k_1 \gg k_{-1} + k$$
$$GDP + Pi \quad GTP$$

V. THE G SHUTTLE MODEL

$$\begin{array}{l} HR + G \cdot GDP \rightleftharpoons HR \cdot G \cdot GDP \\ HR \cdot G \cdot GDP + GTP \rightleftharpoons HR \cdot G \cdot GTP + GDP \\ HR \cdot G' \cdot GTP \longrightarrow HR + G' \cdot GTP \\ G' \cdot GTP + C \longrightarrow G' \cdot GTP \cdot C' \\ C' \cdot G' \cdot GTP \longrightarrow C + G \cdot GDP + Pi \end{array} \left. \begin{array}{l} \\ \\ \\ \\ \end{array} \right\} \begin{array}{l} E = GC \\[1em] k_{on} \\[1em] k_{off} \end{array}$$

Fig. 2: The various molecular models which can describe the coupling between receptor and adenylate cyclase. In models I through V, E depicts the complex N·C.

Among the five families of models depicted in Figure 2, only model IV accounts for the basic findings: (a) Non-cooperative ligand binding to the receptor; (b) Strictly first-order kinetics of adenylate cyclase activation when the enzyme system is challenged with a non-hydrolyzable analog of GTP. This is true for a wide range of experimental conditions. (c) A linear dependence of the rate constant of activation on the concentration of receptors. All other models predict complex ligand binding curves, non-first-order kinetics of activation and a non-linear dependence of the rate on receptor concentration. A detailed summary of the arguments is given in a few recent publications.

8. *Desensitization of β-receptor dependent adenylate cyclase*

The incubation of cells possessing β-adrenergic receptors with agonists leads to a decrease in receptor response. This feature of receptor behavior is common to many receptors and is known as receptor desensitization or subsensitivity. In these studies it was observed that upon repeated β-receptor stimulation, a decreased response of the adenylate cyclase is obtained. A more detailed kinetic analysis of the mechanism of β-receptor desensitization of human astrocytoma cells suggests that the phenomenon of desensitization involves two consecutive steps: (i) a rapid uncoupling of the β-receptors from cyclase, namely, the loss of hormonal stimulation of adenylate cyclase, followed by (ii) loss of β-adrenergic receptors from the cell surface.

The initial event of the loss of receptor to enzyme coupling is reversible, while the loss of β-receptors from the surface of the cells is not. The disappearance of receptors from the frog erythrocyte surface was reported to occur together with the increase of β-receptors in the cytosol, hence suggesting internalization, as was found for polypeptide hormones such as insulin. The reappearance of β-receptors on the cell surface requires the synthesis of new receptors. It is not yet clear whether the internalized receptors are destroyed or recycled. The two stage process of receptor desensitization can be summarized by the scheme:

$$R \underset{k_2}{\overset{k_1}{\rightleftharpoons}} R' \longrightarrow R'' \tag{3}$$

where R is the free unbound receptor, R' is the agonist bound receptor in the uncoupled state, and R" is the internalized receptor which is no longer accessible to ligands. The establishment of the equilibrium $R \rightleftharpoons R'$ seems to be rather fast, although the irreversible process R' to R" conversion is slow and usually spans over a few hours. The R'/R ratio depends on hormone concentration and increases with the efficacy of the agonist. The appearances of cytosolic receptor capable of binding [^{3}H]dihydroalprenolol subsequent to the exposure of the cells to β-agonist, indicates that the binding properties of the receptor are retained. This behavior

may be atypical because the frog erythrocyte may lack degradative
mechanisms that are normally operating in most, if not all, cell
types. In certain cases it was shown that protein synthesis is
required for β-receptor desensitization but its involvement in other
cases was not recognized. It may be that a certain protein component is required for the process of long term desensitization. The
reason that cells differ in their dependence of long term desensitization on protein synthesis may be that the yet unknown protein factor has different life times in different cells.

The uncoupled receptor R' is modified in two identifiable biochemical properties; it is found in a lighter membrane fraction when
the cells are disrupted and the membranes are sedimented in a sucrose
gradient, and it displays a diminished affinity towards the β-agonist
as compared to the non-desensitized receptor. When the agonist is
removed subsequent to the onset of uncoupling but before the receptor
is internalized, the phenomenon is fully reversible: the receptor is
mobilized back to the heavier membrane fraction, it displays normal
affinity to agonist, and it is able to couple and activate adenylate
cyclase.

The achievement of β-receptor desensitization *in vitro* in a
fibroblastic line of normal rat kidney (NRK-S) cells, has recently
been reported. It was shown that the minimal requirements for the
desensitization to occur were the simultaneous presence of β-agonist,
Mg^{2+}, ATP and GTP. The desensitization observed is reflected in the
loss of β-receptor stimulation without loss of receptor sites. In
addition, no loss of activity with PGE_1 or NaF was noted, therefore
verifying that one indeed deals with an agonist-specific desensitization. (App(NH)p was found to be unable to replace ATP in these
experiments, and it was concluded that a phosphorylation reaction
may probably be involved. It has previously been shown that ATP is
required for agonist specific desensitization in another system *in
vitro*. Interestingly, if NRK-S cells are depleted of ATP down to
the micromolar range, desensitization still occurs. Hence, it was
suggested that the desensitization reaction requires a factor in
the phosphorylated form and that none of the functional units of
cyclase are phosphorylated directly. In the intact cell this factor
is still phosphorylated even at low ATP, provided the topography of
the cell remains intact. Presumably, upon rupture, phosphatases
begin to operate and, in order to sustain this factor in the phosphorylated form, significant ATP concentrations are required *in
vitro*. Since the desensitization of the response to β-receptors
is not accompanied by a reduction in the PGE_1 dependent activity
or the NaF stimulated activity, it appears that neither N(G/F) nor
C are modified during the desensitization. It therefore follows
that the receptor must be the entity which is modified. As it does
not appear that phosphorylation is the modification, we have postulated that the change in the properties of the receptor is another

type of covalent modification such as an SH/S-S transition. This modification is catalyzed by a receptor modifying protein which, in itself, is under the control of a phosphorylation-dephosphorylation cycle. The fact that SH to S-S conversion can modulate receptor function, has recently been documented for the morphine receptor. Similarly, the reduction of an S-S bond to SH in the electroplax nicotinic acetylcholine receptor reveals changes in ligand specificity. The reason that an SH to S-S transition has been chosen as the working hypothesis is that different experiments excluded methylation and ADP ribosylation. Interestingly, the modification of the receptor can be reversible, if the agonist is washed from the receptor prior to its internalization.

A number of reports suggest that disruption of cytoskeletal elements by specific drugs inhibits the onset of desensitization. These findings can be interpreted to mean that the biochemical factors catalyzing the desensitization reaction are sub-membranous elements which become dislocated once the internal structure of the cell is disrupted. The other alternative is that the cytoskeleton is involved in some way in the desensitization reaction.

9. *Concluding comments*

Quite a few features of the adenylate cyclase regulatory system still remain largely unknown: (1) The structure of the receptor dependent cyclase regulatory complex is not as yet known. No direct data are yet available on the stoichiometry among the three recognized macromolecules: the receptor, the GTP regulatory unit N(G/F) and the cyclase catalytic moiety. Further progress in this direction will be possible once these macromolecules are purified and, subsequently, be reconsituted into a functional hormone responsive cyclase under controlled experimental conditions. (2) The observations that a few receptors can couple to the same pool of adenylate cyclase strongly suggest that different receptors may possess *common structural domains*. This situation might be taken as an indication that hormone receptors are built like gamma globulins, namely, the receptor consists of a *variable domain* which includes the hormone binding site and a *constant domain* which represents the region of the receptor molecule interacting with the GTP regulatory unit and the catalytic cyclase unit. A similar situation exists for the N(G/F) protein. A variety of N(G/F) proteins can couple to the same catalytic unit C. (3) It is not clear whether all the components of the adenylate cyclase have already been discovered. For example, it has yet to be established whether the GTP regulatory site linked to the cyclase is identical to the site responsible for the hormone dependent GTPase. Purification of the adenylate cyclase components and reconstitution experiments should yield information on that point. So far, the purified N(G/F) proteins were found not to exhibit GTPase activity. (4) How the process of hormone-induced desensitization is linked to the coupling between the hormone receptor and the cyclase catalytic moiety is still an open question. (5) The role of specific membrane lipids as essential co-factors for the receptor to cyclase coupling must still be explored. (6) Whether the fluidity of the membrane matrix is a decisive factor in the mechanism of receptor to enzyme coupling in systems other than the few explored, is not yet known. (7) Connected to the latter point is the question as to which of the hormone receptors known to activate adenylate cyclase are permanently associated with the cyclase and which are separate entities and, therefore, activate the enzyme during transient encounters between the two. In the latter case, the receptors will not be found in the same domains as the enzyme, as was indeed found in one system. For the study of the relative distribution of hormone receptors *vis--à-vis* the cyclase, one will have to develop specific ultrastructural probes for the receptors as well as for the cyclase and the GTP binding protein. (8) The possibility that adenylate cyclase and the GTP regulatory unit, which face the inside of the cell, are anchored to cytoskeletal elements or are freely mobile in the lipid bilayer, must still be explored. Data on the latter point are relevant to the question whether receptor to enzyme coupling is influenced by cytoskeletal elements. These and many other questions regarding receptors which interact with adenylate cyclase are currently being studied.

Comment: Most of the issues discussed in this article are covered in detail in Ref. 1.

References

1. Levitzki, A. (1982) The mode of coupling of adrenergic receptors to adenylate cyclase, in: "Topics in Molecular Pharmacology," A.S.V. Burgen and G.C.K. Roberts, eds., Elsevier/North Holland Publishing Co., pp. 24-62.

BIOCHEMICAL PHARMACOLOGY OF DOPAMINE RECEPTORS

M. Memo, E. Carboni, H. Uzumaki, S. Govoni,
M.O. Carruba, M. Trabucchi, and P.F. Spano

Department of Experimental Pharmacology and Toxicology
University of Cagliari, Department of Pharmacology
Schools of Pharmacy and Medicine, University of Milan
Italy

Over the last few years, new biochemical methodologies have provided direct approaches to extensively investigate central dopaminergic receptors. It has been found that an adenylate cyclase preferentially stimulated by dopamine (DA) is present in several brain dopaminergic areas (16, 20, 35, 37). This finding first allowed a direct pharmacological characterization of compounds with agonist or antagonist properties at the DA receptor level and suggested that physiological responses following interaction between DA and its own receptor are mediated by activation of adenylate cyclase. However, the inhibitory potency of various antipsychotic drugs does not fully correlate with their behavioral and clinical efficacy (33), Moreover, we found that sulpiride and other substituted benzamides, which behave as strong DA antagonists in various animal tests and in clinical applications, fail to inhibit, in contrast to classical neuroleptics, the formation of cyclic AMP elicited by DA or apomorphine both in vitro and in vivo (37, 39, 48). In addition, we observed that some ergot derivatives, endowed with strong DA-mimetic properties in animals and in man, do not stimulate DA-sensitive adenylate cyclase activity in brain homogenates (37, 50). However, we and others have shown that substituted benzamides and dopaminergic ergot derivatives are capable of interacting specifically with central and peripheral DA receptors (15, 37, 39).

These studies provided the basis for a new and useful paradigm. Accordingly, DA receptors associated with adenylate cyclase stimulation have been defined as D_1 while those unassociated have been named D_2 (22, 42). Recently, different groups of investigators suggested that stimulation of D_2 receptor inhibits neurotransmitter-

and neurohormone-stimulated adenylate cyclase (5, 10, 29, 46). In
particular, activation of adenylate cyclase by isoproterenol in
intermediate pituitary and by vasoactive intestinal peptide in anterior pituitary is inhibited by DA in a noncompetitive manner. It
is of interest that dopaminergic ergot alkaloids mimic and sulpiride
competitively blocks the effect of DA on D_2 receptors. Finally, we
and others have provided evidence that a subset of DA D_2 receptors
might be independent of the adenylate cyclase moiety (6, 27, 51).

In this chapter we shall briefly review the biochemical pharmacology of the two classes of DA receptor interacting drugs, namely,
substituted benzamides and dopaminergic ergot derivatives, that have
been instrumental in the elaboration of a new paradigm of DA receptor
organization.

SULPIRIDE AND OTHER BENZAMIDES

Sulpiride is a substituted benzamide that, similar to metoclo-
pramide, possesses a pronounced antiemetic activity. Sulpiride and
other related compounds have been shown to be useful for therapy of
several forms of psychoses, neurotic depression, and other mental
disturbances. In addition, sulpiride possesses strong inhibitory
effects on apomorphine-induced vomiting in dogs and is a potent
stimulant of prolactin release both in laboratory animals and man.
Like classical antipsychotic drugs, sulpiride is able to inhibit
climbing behavior in mice, and elicits accumulation of dihydroxy-
phenylacetic acid (DOPAC) and homovanillic acid (HVA) in DA-reached
brain areas of laboratory animals. However, sulpiride differs from
typical neuroleptics since the increase of brain DOPAC and HVA is
not associated with muscle rigidity or other obvious neurological
changes. Moreover, this drug is a very weak blocker of apomorphine-
or amphetamine-induced stereotyped behavior and produces only minor
extrapyramidal side effects both in animals and man (see reference
61 for a review). Since a preferential distribution of sulpiride
in selected areas of the brain may only partially explain the
peculiar pharmacological profile of this drug, it was tempting to
hypothesize that sulpiride interacts with central DA receptors in
a way basically different from that of classic antidopaminergic
drugs. Therefore, we extensively investigated the interaction of
sulpiride and other substituted benzamides with DA and other neurotransmitter receptors at the molecular level. We found (48) that
sulpiride and other substituted benzamides endowed with antidopaminergic properties, up to a concentration of 10^{-4}M, failed to
inhibit dopamine- or apomorphine-stimulated adenylate cyclase activity either in striatal or nucleus accumbens homogenates (Table 1).
In addition, the failure of substituted benzamides to inhibit the
stimulation of cyclic AMP formation elicited either by submaximal
or supramaximal concentrations of DA was also observed in homogenates of tuberculum olfactorium, substantia nigra, retina, and median
eminence (data not shown).

Table 1. Effect of sulpiride on DA- and apomorphine-stimulated adenylate cyclase activity in homogenates of rat striatum and nucleus accumbens.

Drug	Striatum	Nucleus accumbens
-	204±15	235±17
DA	395±17*	420±31*
Apomorphine	315±21*	348±28*
Sulpiride	190±14	219±23
Sulpiride + DA	360±25*	385±28*
Sulpiride + Apomorphine	300±18*	331±17*

*p < 0.01 if compared to control values.
DA and apomorphine were added at the final concentration of 10 μm; sulpiride at the concentration of 0.1 mM. Values are expressed as pmoles/mg prot/min ± SEM.

Complementary experiments were also performed in our laboratory to assess the effect of sulpiride and its analogs on DA-stimulated formation of cyclic AMP in intact tissue slices (41). This experimental model, although more complex and with its own intrinsic limitations, perhaps more rigorously reproduces the conditions of the physiological environment where drugs and receptors interact with each others.

In these experimental conditions, (-)sulpiride, (+)sulpiride, sultopride, and tiapride were not able to either modify basal adenylate cyclase activity or inhibited the accumulation of cyclic AMP elicited by DA. In contrast, fluphenazine was able to totally counteract the increase of cyclic AMP induced by the inclusion of DA to the incubation medium (41) (data not shown). The results reported so far needed to be corroborated by in vivo experiments in order to rule out that antidopaminergic profile of sulpiride and its congeners could be due to an in vivo formation of active metabolites. For this purpose, the rats were pretreated with supramaximal doses of sulpiride or haloperidol and then a challenge dose of apomorphine was administered 10 min before killing the animals by microwave radiation. It is evident from the data reported in Table 2 that sulpiride, at a dose which produces the maximum effect on DOPAC accumulation, fails to antagonize the increase of striatal cyclic AMP elicited by acute administration of apomorphine. On the contrary, the effect of apomorphine on striatal cyclic AMP content is completely abolished in animals pretreated with haloperidol.

Table 2.   Striatal cyclic AMP content after treatment with
           different DA-agonists and antagonists.

Treatment	Cyclic AMP (pmoles/mg prot ± S.E.M.)	Increase %
Saline	5.5 ± 0.4	-
Apomorphine	8.5 ± 0.6*	54
Haloperidol	5.4 ± 0.4	-
Sulpiride	5.8 ± 0.5	-
Haloperidol + Apomorphine	5.0 ± 0.3	-
Sulpiride + Apomorphine	8.1 ± 0.6*	47

*$p < 0.001$ when compared to saline-treated group.
Sulpiride (50 mg/kg) and haloperidol (1 mg/kg) were given i.p.
3 hr before killing. Apomorphine (2.5 mg/kg) was given 10 min
before killing. Values are the mean of 20 determinations.

    In order to investigate whether sulpiride and other substi-
tuted benzamides could directly interact with central DA receptors,
the effect of these drugs was tested on the specific binding of ^{3}H-
ligands for DA receptors.

    As shown in Table 3 and previously reported in detail else-
where (37, 38), sulpiride and related compounds such as sultopride
and tiapride, are good competitors for ^{3}H-haloperidol binding sites
in rat striatal membrane preparations. Most interestingly, the
interaction of substituted benzamides with the specific binding
sites labeled by tritiated haloperidol appears to be stereospecific since (+)sulpiride, the pharmacologically inactive isomer of
the drug, is devoid of displacing properties. These results have
been confirmed using ^{3}H-spiroperiodol (45). In addition, we
observed that none of substituted benzamide used in our study in-
teracted with ^{3}H-ligands for central muscarinic or GABAergic receptors (37). More recently, the high degree of selectivity of
sulpiride for central DA recognition sites has been demonstrated
further by studies using other ligand binding assays for histamine,
serotonin, and beta-adrenergic receptors (18). It is noteworthy
that sulpiride and other benzamides interact with DA receptors, as
labeled by ^{3}H-spiroperiodol, in an exclusively sodium dependent
manner in contrast to other neuroleptic drugs (45). Similarly, the
binding of ^{3}H-sulpiride appears to be virtually undetectable in the

Table 3.   Drug competition for ^{3}H-haloperidol-specific binding
in membranes prepared from rat striatum.

Drug	Inhibition of ^{3}H-haloperidol binding IC_{50}(nM)
Haloperidol	2.5
Fluphenazine	5
(-)Sulpiride	90
Sultopride	85
Tiapride	100
Dopamine	1000
(+)Sulpiride	10000

Results are the mean of 4 separate experiments, each using seven or
more concentrations of drugs, with sample run in triplicate.   SEM
were less than 5% of the means.

absence of sodium ions while the specific binding of ^{3}H-spiroperone
is only partially sodium dependent (2, 47).

Since in the conventional method used to measure DA-stimulated
adenylyl cyclase activity in cell-free preparations the incubation
medium does not contain sodium ions, it was imperative to test the
effect of sulpiride on this system in the presence of sodium.   We
found that the inclusion of sodium ions in the standard cyclase
assay buffer has no effect on the failure of sulpiride and other
benzamide derivatives to inhibit DA-stimulated adenylate cyclase
activity in striatal homogenates (data not shown).

One alternative explanation for the lack of effect of sulpiride
on DA-stimulated adenylate cyclase activity argues that this drug
might not achieve a sufficiently high degree of membrane penetration
to act as a dopamine antagonist (52).   Thus only drugs with a sufficiently high oil/water partition coefficient will function as
dopamine antagonists in this system.   We have measured the oil/water
partition coefficient of a number of compounds endowed with antidopaminergic properties (44).   No correlation whatsoever could be
found in our experiments between lipid solubility and the ability
to antagonize DA-stimulated adenylyl cyclase activity (data not
shown).

Table 4.   Properties of ^{3}H(-)sulpiride binding in striatal
membrane preparations.

Binding parameters	
1. High affinity	K_D = 5.2 nM
2. Saturable	Hill coefficient = 0.94
3. Definite number of sites	B_{max}= 295 fmol/mg prot
4. Same affinities for equlib-rium and kinetic constants	K_D (equilibrium) = 5.2 nM K_D (kinetic) = 4.5 nM
5. Sensitivity to ions	Completely Na^+ dependent
6. Sensitivity to temperature	Maximum binding capacity at 8°C
7. Sensitivity to GTP	Yes, decreased affinity for DA agonists
8. Pharmacology	Selectively displaced by DA agonists and antagonists

CHARACTERIZATION OF ^{3}H(-)SULPIRIDE SPECIFIC BINDING TO RAT
CEREBRAL MEMBRANES

     (-)Sulpiride is an antipsychotic drug endowed with dopamine
antagonist properties. The failure of (-)sulpiride to inhibit neo-
striatal DA-stimulated adenylate cyclase suggested that this compound
acts as a selective antagonist at those DA receptors functioning independently from adenylate cyclase stimulation (D_2). Recently, we
used a novel synthetized ^{3}H(-)sulpiride endowed with very high specific activity (72 Ci/mmol), and characterized the stereospecific
binding of this compound in various rat brain areas (2). The properties of ^{3}H(-)sulpiride in rat striatal membranes are summarized
in Table 4. In accord with previous results (26), the stereospecific binding appears saturable, temperature dependent, stereospecific, and showed very high association and dissociation rate values.
The stereospecific binding was defined as the difference in binding
between samples incubated with 5 x 10^{-7} M (+)sulpiride and samples
incubated with 5 x 10^{-7} M (-)sulpiride. K_D values calculated by
saturation isotherms and by kinetic rate constant measurements were
essentially similar (5.2 nM and 4.5 nM, respectively). Binding
capacity in various brain regions was at a maximum in striatum and
decreased according to the following order: hypothalamus, substantia
nigra, tuberculum olfactorium, and pituitary. No specific
binding was detectable in cerebellum. Subcellular localization
studies revealed that crude synaptic membrane fractions possess the
highest density of ^{3}H(-)sulpiride binding sites. Ions and GTP

altered ^{3}H(-) sulpiride binding characteristics in DA brain regions
where specific binding sites were found. In particular, measurement of ^{3}H(-)sulpiride binding was selectively Na$^+$-dependent,
reaching the maximal value with 120 mM NaCl. GTP did not change
either B_{max} or K_D values of ^{3}H(-)sulpiride specific binding. However, the IC_{50} of DA increased by 5 to 6 times when incubation was
conducted in the presence of 50 μM GTP. We also tested the potency
of different DA agonists and antagonists in displacing ^{3}H(-)sulpiride
binding, and found that all of them, including cis-flupentixol and
SKF 38393, were active in this test. The results suggested that
SKF 38393 and Cis-Flupentixol may interact with DA receptor subtypes
other than D_1. In this regard ^{3}H(-)sulpiride may be a useful tool
for the characterization and localization of different D_2 receptor
subtypes.

DOPAMINERGIC ERGOT DERIVATIVES

Since the classical studies of H. H. Dale in 1906 (8), the
primary site of action of ergot alkaloids has been regarded to be
at the receptor level. The ergot alkaloids were the first adrenergic blocking agents to be discovered, and the hydrogenated derivatives are among the most potent α-adrenergic antagonists known.
However, the emphasis on their α-adrenergic blocking activity has
frequently caused other important pharmacological properties to be
overlooked. Side effects, indicatives of peculiar target sites,
prevent the administration of doses that could produce more than
the minimal adrenergic blockade in man. The dose of dihydroergo-
toxine in man is, for instance, strictly limited by the production
of nausea and vomiting. In fact, all the peptide ergot alkaloids
have potent emetic properties due to stimulation of the chemore-
ceptor trigger zone (CTZ) in the medulla oblongata. Other important
actions of many ergot derivatives are a profound inhibition of
prolactin secretion in vivo and in isolated pituitary preparation
in vitro (28, 36).

The effects on the CTZ and mammotrophs are characteristic of
drugs endowed with dopamine mimetic activity and have stimulated
the search for other dopamine-receptor mediated phenomena. Thus a
large body of data has been provided suggesting that various ergot
alkaloids may indeed preferentially stimulate dopamine receptors in
the mammalian brain and that this effect could contribute to the
marked behavioral and endocrinological actions elicited by these
drugs (13, 14).

On the basis of these findings, a certain number of ergot compounds have been defined as dopaminergic ergot derivatives and have
been found to have clinical applications as antiparkinsonian
agents, as inhibitors of prolactin secretion in human galactorrhea
and gonadal dysfunction, and as inhibitors of growth hormone

Table 5. Effects of various DA-mimetic compounds on DA-stimulated
          adenylate cyclase activity in homogenate of rat striatal
          membranes.

Drug	Maximum stimulation[a] (% dopamine)	EC_{50}[b] (μM)	IC_{50}[c] (μM)
Dopamine	100	2.0	–
Epinine	100	1.5	–
ADTN	100	2.1	–
Apomorphine	56	1.9	30
Ergometrine	47	2.2	0.9
Lisuride	–	–	0.1
Bromocriptine	–	–	0.3
Lergotrile	–	–	0.3
DHE	–	–	3.0

[a]Maximum stimulation above basal cyclic AMP production is expressed
as a percentage of that obtained with 100 μM DA.

[b]EC_{50} is the drug concentration needed to produce 50% of the
maximum effect.

[c]IC_{50} is the drug concentration required to inhibit the maximum
stimulation induced by DA by 50%. Each IC_{50} value was determined
from a logprobit using three or more concentrations of drug.

secretion in acromegalic patients. In addition, these drugs have
been proposed to be useful in geriatry and gerontology. Dopaminergic ergot drugs produce a spectrum of behavioral effects which do
not completely mimic those of apomorphine either qualitatively or
quantitatively. The behavior elicited by ergots and apomorphine
are also differentially sensitive to DA receptor antagonists (13,
36).

Taken together, these differences suggest that at least some
of the dopaminergic ergot derivatives may produce behavioral effects
by an agonistic action on DA receptors different from those activated by apomorphine. In 1976 we observed that one of the most potent
dopaminergic ergot derivatives (49, 50), namely bromocriptine, in
contrast to the classical DA-agonist apomorphine and DA itself, was
not capable of stimulating the formation of cyclic AMP in homogenates of rat striatum and nucleus accumbens. Rather, bromocriptine was almost as active as haloperidol in inhibiting the stimulation of adenylate cyclase activity induced by DA.

Later, similar results were reported by Kebabian et al. (21) for another potent dopaminergic ergot derivative, lergotrile. Thereafter, various ergot compounds were examined with essentially identical findings (38, 43). The results obtained in our laboratories are summarized in Table 5. With the exception of ergometrine, which, like apomorphine, behaves as a partial agonist, no agonist activity could be demonstrated for bromocriptine, dihydroergocryptine (DHE), and lisuride. In contrast, these drugs behave as inhibitors of the stimulation produced by dopamine, that is, they appear to be antagonists. Moreover, it is interesting to note that the antagonistic properties exhibited in vitro by ergot derivatives are basically different from those of classical DA-antagonists. As shown in Figure 1, bromocriptine, in contrast to fluphenazine, inhibits DA-stimulated adenylate cyclase in a noncompetitive way. A similar pattern of kinetic behavior applies to other dopaminergic ergot derivatives such as lisuride, lergotrile, and DHE (data not shown).

The lack of adenylate cyclase activity stimulation in vitro associated with DA mimetic effects in vivo could be interpreted as the dopaminergic ergot derivatives' exertion of their pharmacological actions through active metabolites. However this possibility may be ruled out since the studies of Keller et al. (23) who demonstrated that at least bromocriptine, lisuride, and lergotrile elicit dopaminergic behavioral responses as parent compounds.

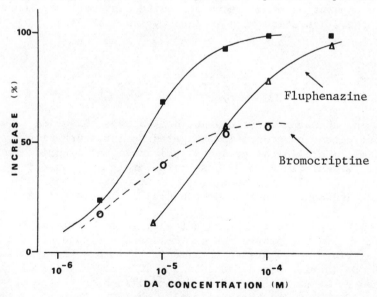

Fig. 1. Inhibition by fluphenazine and bromocriptine of adenylate cyclase stimulation induced by different concentrations of dopamine. Fluphenazine and bromocriptine were added at the concentration of 5×10^{-5} M.

Table 6.   Properties of ^{3}H-lisuride binding in rat striatal
membrane preparations.

Binding parameters	
1. High affinity	K_D = 1.8 nM
2. Saturable	Yes, Hill coefficient = 0.96
3. Definite number of sites	B_{max} (Striatum) = 723 fmole/mg of protein
4. Same affinities for equilibrium and kinetic constants	K_D(kinetic) = 1.6 nM K_D(equilibrium) = 1.8 nM
5. Sensitivity to ions	Partially Na^+ dependent
6. Sensitivity to GTP	No
7. Pharmacology	Selectively displaced by dopaminergic agonists and antagonists

Thus, the most simple hypothesis which can be drawn is that dopaminergic ergot derivatives preferentially stimulate, at low concentration, a population of DA receptors quite distinct from those triggering the stimulation of adenylate cyclase activity.

CHARACTERIZATION OF ^{3}H-LISURIDE SPECIFIC BINDING TO RAT STRIATAL
MEMBRANES

Among the dopaminergic ergot derivatives, lisuride is one of
the most potent compounds investigated so far. Behavioral, pharmacological, and clinical evidence indicates that lisuride can indeed
directly stimulate DA receptors in the brain and anterior pituitary
(3, 9, 17, 24, 31). Furthermore, in vitro experiments using ^{3}H-lisuride indicate that this compound binds to central DA receptors
(12, 51). However, lisuride, unlike apomorphine, does not stimulate
DA-sensitive adenylate cyclase activity in rat striatal homogenates
(30, 32, 41). The properties of ^{3}H-lisuride binding in rat
striatal preparation are summarized in Table 6. The specific binding of ^{3}H-lisuride was defined by 5 x 10^{-7} M d-butaclamol and
appeared to be saturable and stereospecific both in the presence or
absence of NaCl. The presence of 120 mM NaCl, however, increased
the B_{max} by 40% without changing the K_D. This finding raises the
possibility that ^{3}H-lisuride binds more than one site in striatal
membrane preparations. In order to further investigate how dopaminergic ergots interact with DA receptors, we studied the effect

Table 7. Effect of GTP on binding properties of ^{3}H-lisuride and ^{3}H-apomorphine

	-GTP	+GTP
[3]H-lisuride		
B_{max}	723	754
K_D	1.8	1.7
[3]H-apomorphine		
B_{max}	351	377
K_D	0.9	5.6*

*$p < 0.01$ with respect to the corresponding values in the absence of GTP.
B_{max} values are expressed as fmol/mg prot; K_D values as nM.

Table 8. Competition of various dopaminergic drugs for ^{3}H-spiro-peridol binding in presence or absence of GTP.

Drug	-GTP	+GTP
DA	2000	8000*
ADTN	300	1180*
Apomorphine	110	460*
Bromocriptine	7.6	7.7
DHE	7.8	7.8
Lisuride	6.5	6.9

*$p < 0.01$ with respect to the corresponding values in the absence of GTP.
Results are the mean of 4 separate experiments, each using three or more concentrations of drugs, with samples run in triplicate. SEM were less than 5% of the means.

of GTP on ^{3}H-lisuride binding as compared to ^{3}H-apomorphine binding, and on the displacement of ^{3}H-spiroperidol binding by classical DA mimetic drugs such as apomorphine and amino-6,7-dihydroxy-1,2,3,4-

tetrahydronapthalene (ADTN) as compared to dopaminergic ergot derivatives, such as bromocriptine, lisuride and DHE. Guanine nucleotides have been shown to regulate the binding of several hormone
and neurotransmitter receptors, especially those associated with
adenylate cyclase. As summarized in Tables 7 and 8, GTP at the concentration of 50 µM did not affect the ^{3}H-lisuride binding characteristics, but the affinity of ^{3}H-apomorphine binding was decreased to
about one half in the presence of GTP. When we examined the effect
of GTP on ^{3}H-lisuride binding in the presence of NaCl, it appeared
that neither the K_D nor the B_{max} of ^{3}H-lisuride binding were modified by GTP (51). Similarly, no effect of GTP could be detected on
the inhibition of ^{3}H-spiroperidol binding by lisuride, by bromocriptine, and DHE. On the other hand, the potency of DA, apomorphine,
and ADTN in inhibiting ^{3}H-spiroperidol binding was reduced by three-
fold. According to these findings, we suggest that lisuride and
the other dopaminergic ergot derivatives primarily bind to a DA
receptor not linked to stimulatory or inhibitory nucleotide units,
and therefore to adenylate cyclase moiety.

ORGANIZATION OF DA RECEPTORS

    Receptor function is geared to detect specific chemical signals
that reach the outer surface of the neurons where the receptor is
located, and to cause the internalization of these signals, transforming them into metabolic stimuli. Currently, DA receptors are
considered to be important regulatory sites of pre- and postsynaptic function: they adapt to environmental changes by virtue of
their plasticity which depends on a certain functional flexibility
of their organization. The supramolecular complex of DA receptors
resembles that of receptors for other neurotransmitters, and it is
believed to include three basic units: the detector or recognition
site for DA, the coupler which may be a protein that couples the
recognition site with the transducer, and the transducer (see
references 7 and 11 for reviews).

    It is generally agreed that in corpus striatum, nucleus accum-
bens, and other DA-reached areas of the brain, the post-synaptic
recognition sites for DA are, at least in part, functionally linked
to adenylate cyclase in a stimulatory way (1, 4, 19, 34). This
link appears to be regulated by an intrinsic protein, termed G/F
protein. The G/F protein consists of several subunits which have
binding sites for GTP, GDP, and NaF. When the DA recognition sites
are occupied by an agonist which activates adenylate cyclase, the
GDP bound to G/F protein is released while the binding of GTP to
this protein is facilitated. This process is considered to be the
rate limiting step for adenylate cyclase activation. Besides G/F
protein, many other components seem to modulate the amplification
of DA signal. In this regard, the Ca^{++}-binding protein Calmodulin
has been shown to be involved in the process of desensitizating

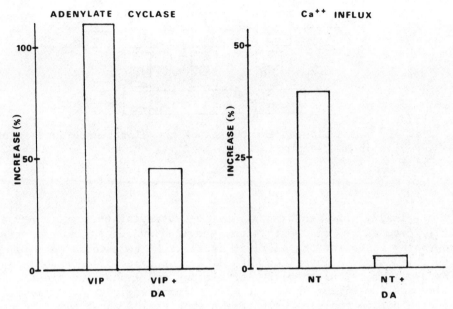

Fig. 2. Dopamine inhibition of intracellular modifications induced
by VIP and neurotensin in rat pituitary. Adenylate cyclase activity
was measured in homogenates obtained from male rat pituitary. Basal
activity was 43±4 pmol of cyclic AMP formed/mg prot/min. DA alone
did not modify basal enzyme activity. Ca++ influx was measured
using a standard procedure with-^{45}Ca++ in 30 min incubated emipi-
tuitary. Basal ^{45}Ca++ incorporation into the tissue was 8.3
pmol/gland. DA alone did not modify basal Ca++ influx. All the
peptides and DA were added at the concentration of 10-^{7}M.

DA receptors (25). Recently, various groups of investigators have
independently provided evidence that DA and dopaminergic agonists
may also inhibit, under certain conditions, neurotransmitter- or
neurohormone-stimulated adenylate cyclase activity.

Kebabian and co-workers (5) have demonstrated an adenylate
cyclase preferentially stimulated by isoproterenol in the anterior
pituitary lobe. The stimulated cyclase activity is inhibited by
β-adrenergic blocking agents in a competitive way and by DA and dopaminergic ergot derivatives in a noncompetitive way. Interestingly,
sulpiride stereospecifically antagonized DA inhibition of isopro-
terenol-stimulated adenylate cyclase. Quite independently, Costa
and co-workers (29) have provided evidence that adenylate cyclase
activity stimulated by vasoactive intestinal peptide (VIP) in
anterior pituitary is inhibited by DA and dopaminergic agonists in
a noncompetitive manner. In this model also, sulpiride stereospecifically antagonized the inhibitory effect of DA and dopaminergic compounds on VIP-stimulated adenylate cyclase activity.

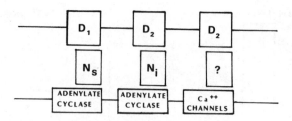

Fig. 3. Schematic representation of the different organization
of DA receptors.

Finally, the Kebabian group (46), studying the cyclic AMP efflux from slices of neostriatum, provided evidences indicating
opposing roles for the distinct types of DA receptors. According
to this data, DA interacts with different types of receptors in
neostriatum, simultaneously stimulating and inhibiting the formation and the efflux of cyclic AMP. Dopaminergic ergot derivatives
and sulpiride could be selectively agonist and antagonist, respectively, of the dopamine receptors which mediate the inhibition of
cyclic AMP formation. DA inhibition of adenylate cyclase appears
to be dependent of the presence of GTP in the incubation medium.
These results suggest that in this case the recognition site for
DA is coupled with a peculiar type of G/F protein that causes an
inhibition of adenylate cyclase activity.

More recently, we have provided evidence that by a receptor
interaction in anterior pituitary DA and dopaminergic agonists,
modulate neurotensin (NT)-induced increases in PRL release (27).
This effect appears to be completely independent of the adenylate
cyclase system, but associated to Ca^{++} channels. We found that
neurotensin increases PRL release by a mechanism which involves an
enhanced entry of calcium into the cells. NT-induced increase in
Ca^{++} influx is specifically antagonized by DA and dopaminergic
drugs such as apomorphine and bromocriptine (Figure 2). Furthermore, (-)sulpiride, but not (+)sulpiride, counteracts this DA effect (data not shown).

Taking into account the intracellular modifications following
the interaction between DA and its own recognition sites, it can
be suggested that in the cerebral tissue the recognition sites for
DA are coupled to various kinds of amplifier systems with different
molecular mechanisms. This variety may represent the molecular
basis for the diversity in pharmacological and biochemical profiles
of different DA receptors located in various brain areas and peripheral tissue. On the other hand, the specificity of some classes of
drugs in interacting with peculiar types of DA receptors (substituted benzamides and DAergic ergot derivatives) indicate that the

DA recognition site containing protein is not always identical in the DA receptor complex. According to the pharmacological data, at least two types of DA recognition sites, originally termed D_1 and D_2 (42), can be characterized. Further studies suggested that the different recognition sites for DA are indeed linked with different couplers and transducers. Particularly, the transducer can transform the chemical sign encoded in DA into an inhibitory modulation of a specific ion flux through the post-synaptic membrane (Ca^{++} channel) or into an enzyme activity regulation (cyclic AMP formation or inhibition). A schematic representation of the different organizations of DA receptors is depicted in Figure 3.

According to the data presented and discussed in the text, dopamine D_2-recognition sites might exist in two different receptor complexes. In one case they would functionally inhibit adenylate cyclase activity by coupling with a peculiar type of G/F protein (Ni). In the other, they are possibly independent from adenylate cyclase moiety but associated to Ca^{++} channel. The molecular nature of the coupling device in the latter has yet to be elucidated. However, D_1 recognition sites appear to be linked to adenylate cyclase through a G/F protein (Ns) in a stimulatory way.

ACKNOWLEDGMENTS

The authors wish to thank Mrs. A. Ciuti for her assistance in preparing the manuscript.

REFERENCES

1. Brown, J.N., and Makman, M.H., Stimulation by dopamine of adenylate cyclase in retinal homogenates and of adenosine 3',5'-cyclic AMP formation in intact retina, Proc. Natl. Acad. Sci. USA, 69: 539-543, 1972.

2. Carboni, E., Memo, M., Trabucchi, M., and Spano, P.F., Temperature-dependency of $^3H(-)$sulpiride binding in various areas of rat brain (submitted for publication).

3. Carruba, M.O., Ricciardi, S., Müller, E.E., and Mantegazza, P., Anorectic effect of lisuride and other ergot derivatives in the rat, Eur. J. Pharmacol., 64: 133-141, 1980.

4. Clement-Cormier, Y.C., and Robinson, G.A., Adenylate cyclase from various dopaminergic areas of the brain and the action of antipsychotic drugs, Biochem. Pharmacol., 26: 1719-1722, 1977.

5. Cote, T.E., Orewe, C.W., and Kebabian, J.W., Stimulation of a D_2-dopamine receptor in the intermediate lobe of rat pituitary gland decreases the responsiveness of the β-adrenoreceptor: biochemical mechanism, Endocrinology, 108: 420-426, 1981.

6. Creese, I., Usdin, T.B., and Snyder, S.H., Dopamine receptor binding regulated by guanine nucleotides, Mol. Pharmacol., 16: 69-76, 1979.

7. Cuatrecasas, P., Membrane receptors, Annu. Rev. Biochem., 43: 169-214, 1974.

8. Dale, H.H., On some physiological actions of ergot, J. Physiol. (London), 34: 163-206, 1906.

9. Da Prada, M., Bonetti, E.P., and Keller, H.H., Induction of mounting behavior in female and male rats by lisuride, Neurosci. Lett., 6: 349-353, 1977.

10. De Camilli, P., Macconi, D., and Spada, A., Dopamine inhibits adenylate cyclase in human prolacting-secreting pituitary adenomas, Nature (London), 278: 252-254, 1979.

11. De Haen, C., The non-stoichiometric floating receptor model for hormone-sensitive adenylate cyclase, J. Theor. Biol., 58: 383-400, 1976.

12. Fujita, N., Saito, K., Yonehara, N., and Yoshida, H., Lisuride inhibits ^{3}H-spiroperidol binding to membranes isolated from striatum, Neuropharmacol., 17: 1089-1091, 1978.

13. Fuxe, K., and Calne, D.B., (eds), Dopaminergic Ergot Derivatives and Motor Function, Pergamon Press, Oxford, 1979.

14. Goldstein, M., Calne, D.B., Lieberman, A., and Thorner, M.O., (eds.), Advances in Biochemical Psychopharmacology, Vol. 24, Raven Press, New York, 1980.

15. Goldstein, M., Lew, J.Y., Hata, F., and Lieberman, A., Binding interactions of ergot alkaloids with monoaminergic receptors in the brain, Gerontology, 24: 76-85, 1978.

16. Horn, A.S., Cuello, A.C., and Miller, R.J., Dopamine in the mesolimbic system of rat brain: endogeneous levels and the effect of drugs on the up-take mechanisms and stimulation of adenylate cyclase, J. Neurochem., 22: 265-270, 1974.

17. Horowski, R., Differences in the dopaminergic effects of the ergot derivatives bromocriptine, lisurdie, and d-LSD as compared with apomorphine Eur. J. Pharmacol., 51: 157-166, 1978.

18. Hyttel, J., Characterization of ^{3}H-GABA receptor binding to rat brain synaptosomal membranes: effect of non-GABAergic compounds, Psychopharmacol., 65: 211-214, 1979.

19. Kebabian, J.W., and Greengard, P., Dopamine-sensitive adenylate cyclase: possible role in synaptic transmission, Science, 174: 1346-1349, 1971.

20. Kebabian, J.W., Petzold, G.L., and Greengard, P., Dopamine-sensitive adenylate cyclase in caudate nucleus of rat brain and its similarity to the "dopamine receptor," Proc. Natl. Acad. Sci. USA, 69: 2145-2149, 1972.

21. Kebabian J.W., Calne, D.B., and Kebabian, P.R., Lergotrile mesylate: an in vivo dopamine agonist which blocks dopamine receptors in vitro, Comm. Psychopharmacol., 1: 311-318, 1977.

22. Kebabian, J.W., and Calne, D.B., Multiple receptors for dopamine, Nature, 277: 93-96, 1979.
23. Keller, H.H., and Da Prada, M., Central dopamine agonistic activity and microsomal biotransformation of lisuride, lergotrile and bromocriptine, Life Sci., 24: 1211-1222, 1979.
24. Liuzzi, A., Chiodini, P.G., Opizzi, G., Botalla, L., Verde, G., De Stefano, L., Colussi, G., Graf, K.J., and Horowski, R., Lisuride hydrogen maleate: evidence for a long-lasting dopaminergic activity in humans, J. Clin. Endocrinol. Metab., 46: 196-202, 1978.
25. Memo, M., Lovenberg, W., and Hanbauer, I., Agonist-induced sub-sensitivity of adenylate cyclase coupled with a dopamine receptor in slices from rat corpus striatum, Proc. Natl. Acad. Sci. USA, 79: 4456-4460, 1982.
26. Memo, M., Govoni, S., Carboni, E., Trabucchi, M., and Spano, P.F., Characterization of stereospecific binding of ^{3}H-(-)sulpiride, a selective antagonist at dopamine-D_2 recep-tors, in rat CNS, Pharmacol. Res. Comm., 15: 191-199, 1983.
27. Memo, M. Carboni, E., and Spano, P.F., Dopamine inhibits neuro-tensin-induced prolactin release by interacting with calcium channel rather than adenylate cyclase systems, Soc. Neurosci. Abstr., 1983, in press.
28. Nickerson, M., The pharmacology of adrenergic blockade, Pharmacol. Rev., 1: 27-101, 1969.
29. Onali, F.L., Schwartz, J.P., and Costa, E., Dopaminergic modula-tion of adenylate cyclase stimulation by vasoactive intes-tinal peptide in anterior pituitary, Proc. Natl. Acad. Sci. USA, 78: 6531-6534, 1981.
30. Pieri, L., Keller, H.H., Burkard, W., and Da Prada, M., Effects of lisuride and LSD on cerebral monoamine systems and hallucinosis, Nature, 272: 278-280, 1978.
31. Pieri, M., Schaffner, R., Pieri, L., Da Prada, M., and Haefely, W., Turning in MFB-lesioned rats and antagonism on neuro-leptic-induced catalepsy after lisuride and LSD., Life Sci., 22: 1615-1622, 1978.
32. Saiani, L., Trabucchi, M., Tonon, G.C., and Spano, P.F., Bromo-criptine and lisuride stimulate the accumulation of cyclic AMP in intact slices but not in homogenates of rat neos-triatum., Neurosci. Lett., 14: 31-36, 1979.
33. Seeman, P., Lee, T., Chau-Wong, M., and Wong, K., Antipsychotic drugs doses and neuroleptic/dopamine receptors, Nature (London), 261: 717-719, 1976.
34. Spano, P.F., Di Chiara, G., Tonon, G.C., and Trabucchi, M., A dopamine stimulated adenylate cyclase in rat substantia nigra., J. Neurochem., 27: 1565-1568, 1976.
35. Spano, P.F., Trabucchi, M., and Di Chiara, G., Localization of nigral dopamine-sensitive adenylate cyclase on neurons originating from the corpus striatum, Science, 196: 1343-1345, 1977.

36. Spano, P.F., and Trabucchi, M., (eds.), Ergot Alkaloids, Phar-
    macology, Vol. 16, Suppl., 1, Karger, Basel, 1978.
37. Spano, P.F., Govoni, S., and Trabucchi, M., Studies on Phar-
    macological properties of dopamine receptor in various
    areas of the central nervous system, In: Advances in Bio-
    chemical Psychopharmacology, Raven Press, New York, 19:
    155-165, 1978.
38. Spano, P.F., Frattola, L., Govoni, S., Tonon, G.C., and
    Trabucchi, M., Dopaminergic ergot derivatives: selective
    agonists of a new class of dopamine receptors. In: Dopa-
    minergic Ergot Derivatives and Motor Function, Fuxe, K.,
    and Calne, D.B., (eds.), pp. 159-171, Pergamon Press,
    Oxford, 1979.
39. Spano, P.F., Stefanini, E., Trabucchi, M., and Fresia, P.,
    Stereospecific interaction of sulpiride with striatal and
    nonstriatal dopamine receptors. In: Sulpiride and Other
    Benzamides, Spano, P.F., Trabucchi, M., Corsini, G.U.,
    Gessa, G.L., (eds.), pp. -1-31, IBREF, Milan and Raven
    Press, New York, 1979.
40. Spano, P.F., Trabucchi, M., Corsini, G.U., and Gessa, G.L.,
    (eds.), Sulpiride and Other Benzamides, IBREF, Milan;
    Raven Press, New York, 1979.
41. Spano, P.F., Saiani, L., Memo, M., and Trabucchi, M., Inter-
    action of dopaminergic ergot derivatives with cyclic
    nucleotide system, In: Advances in Biochemical Pharmacol-
    ogy, Raven Press, New York, 23: 95-102, 1980.
42. Spano, P.F., Memo, M., Stefanini, E., Fresia, P., and Trabucchi,
    M., Detection of multiple receptors for dopamine, In:
    Receptor for Neurotransmitters and Peptide Hormones,
    Pepeu, G., Kuhar, M.J., and Enna, S.J., (eds.), pp. 243-
    251, 1980, Raven Press, New York.
43. Spano, P.F., Govoni, S., Uzumaki, H., Bosi, A., Memo, M.,
    Lucchi, L., Carruba, M., and Trabucchi, M., Stimulation
    of D_2-dopamine receptors by dopaminergic ergot alkaloids:
    studies on the mechanism of action, Aging, 23: 165-177,
    1983.
44. Spano, P.F., Carboni, E., Garau, L., Memo, M., Govoni, S., and
    Trabucchi, M., Sulpiride and other benzamides as specific
    antagonists at the D_2 dopamine receptors, In: Receptors
    as Supramolecular Entities, Biggio, G., Costa, E., Gessa,
    G.L., and Spano, P.F., (eds.), Pergamon Press, 1983;
    in press.
45. Stefanini, E., Marchisio, A.M., Devoto, P., Vernaleone, F.,
    Collu, R., and Spano, P.F., Sodium-dependent interaction
    of benzamides with dopamine receptors, Brain Res., 198:
    229-233, 1980.
46. Stoof, J.C., and Kebabian, J.W., Opposing roles for D_1 and D_2
    dopamine receptors in efflux of cyclic AMP from rat
    neostriatum, Nature, 294: 366-368, 1981.

47. Theodorou, A.E., Hall, M.D., Jenner, P., and Marsden, C.D.,
    Cation regulation differentiates specific binding of
    ^{3}H-sulpiride and ^{3}H-spiroperone to rat striatal prepara-
    tions, J. Pharm. Pharmacol., 32: 441-444, 1981.
48. Trabucchi, M., Longoni, R., Fresia, P., and Spano, P.F.,
    Sulpiride a study of the effects on dopamine receptors in
    rat neostriatum and limbic forebrain, Life Sci., 17:
    1551-1556, 1975.
49. Trabucchi, M., Govoni, S., Tonon, G.C., and Spano, P.F.,
    Localization of dopamine receptors in the rat cerebral
    cortex, J. Pharm. Pharmacol., 28: 244-245, 1976.
50. Trabucchi, M., Spano, P.F., Tonon, G.C., and Frattola, L.,
    Effect of bromocriptine on central dopaminergic receptors,
    Life Sci., 19: 225-232, 1976.
51. Uzumaki, H., Govoni, S., Memo, M., Carruba, M., Trabucchi, M.,
    and Spano, P.F., Effects on GTP and sodium on rat striatal
    dopamine receptors labeled by ^{3}H-lisuride, Brain Res.,
    248: 185-187, 1982.
52. Woodruff, G.N., Freedman, S.B., and Poat, J.A., Why does sul-
    piride not block the effect of dopamine on the dopamine-
    sensitive adenylate cyclase?, J. Pharm. Pharmacol., 32:
    802-803, 1980.

BENZODIAZEPINE RECEPTOR LIGANDS WITH

POSITIVE AND NEGATIVE EFFICACY*

Claus Braestrup[1,2], Mogens Nielsen[1] and Tage Honoré[2]

[1]Sct. Hans Mental Hospital, 4000 Roskilde, Denmark
[2]A/S Ferrosan, Sydmarken 5, DK-2860 Soeborg, Denmark

BENZODIAZEPINE RECEPTORS

A major advance in the understanding of the mechanism of action of benzodiazepines and in the understanding of anxiety was the discovery, in 1977, of benzodiazepine (BZ) receptors[1,2] (for a review[3]). Radioactively labelled diazepam binds to a protein in the neuronal plasma membrane and this membrane protein was shown to be embedded in the outer lipid membrane of the cell. Studies with the selective neurotoxic agent kainic acid and on the mutant "nervous" mouse (nr/nr) (in which the Purkinje cells in the cerebellum completely degenerate) located the binding protein to neurons in the central nervous system. High affinity BZ receptors are not found peripherally.

Before any biological entity such as a binding protein can be established as a receptor it is essential to demonstrate a relation between the putative receptor and a characteristic function of the neurotransmitter, hormone, or drug with which the receptor is presumed to interact. This is easy in some peripheral systems where, for eaxample, the presence of a cholinergic agonist at cholinergic receptors in muscles characteristically contratcts the muscle. A functional receptor interaction is less easily demonstrated for benzodiazepines because no measurable direct consequence of their action is known. An indirect approach was adopted. Different benzodiazepines offer a wide range of pharmacological potencies, and it was demonstrated that these potencies correlate well with the

* This paper was also presented at the Soc. Neuroscience Minneapolis, November 1982, and at "CNS Receptors - from Molecular Pharmacology to Behaviour", Strassbourg, September 1982.

strength (K_i values or ED_{50} values) with which the benzodiazepine
binds to the binding protein in vitro[1-3] or in vivo (Fig. 1). Such a
correlation suggests that binding has functional relevance and
indicates that the binding protein is a receptor for benzodiazepines.
The demonstration that BZ receptors and GABA receptors are coupled
at the molecular level (see below) further substantiates the relevance of BZ receptors.

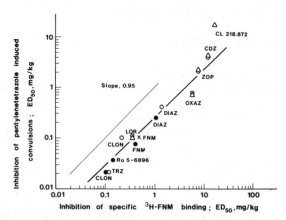

Fig. 1. Correlation between pharmacological effect and benzodiazepine receptor occupancy in vivo for various benzodiazepines,
zoplicone, and CL 218.872 (from ref. 4).

> At ED_{50} the agent inhibits the development
> of clonic convulsions in 50% of mice treated
> subcutaneously with 150 mg/kg of pentylene-
> tetrazol (ordinate) or inhibits by 50%
> specific binding of ^{3}H-flunitrazepam (i.v.)
> in the brain of the living mouse. At ED_{50} an
> agent occupies 50% of the available benzo-
> diazepine receptors (abscissa).
> CDZ = chlordiazepoxide, ZOP = zopiclone,
> OXAZ = oxazepam, MID = midazolam, DIAZ =
> diazepam, LOR = lorazepam, FNM = flunitrazepam,
> CLON = clonazepam, TRZ = triazolam.
> Different symbols denotes different administra-
> tion forms and different investigators (see
> ref. 4).

The strength with which a benzodiazepine binds to the receptors
does not relate in practical terms to the duration of action.
Irrespective of potency, the benzodiazepines shown in Fig. 1 occupy
about 25% of BZ receptors at the dose at which they antagonise
pentylenetetrazol. We still know hardly anything about the percentage of BZ receptors occupied in the human brain during treatment of
anxiety.

The BZ receptor proteins are heterogeneous. There are two classes of receptor, BZ_1 and BZ_2, which are either distinct proteins or two conformational states of one receptor[5-7]. Do these two receptors serve distinct functions? If so, selective agents could be developed and yield new and specific drugs. Preliminary evidence suggests that BZ_2 receptors may be related to anxiety[4]. Receptor occupancy at BZ_2 receptors rather that at BZ_1 receptors seems to determine anti-conflict effects in rats[4] and a regional study of the distribution of BZ_2 (as a percentage of total BZ receptor population) in the monkey brain revealed that BZ_2 receptors mainly reside in limbic brain regions such as the amygdaloid complex, the hippocampal formation, and parts of the prefrontal cortex. These phylogenetically old brain areas have for years been related to anxiety.

BZ receptors, as measured by the total number of receptors, are not particularly abundant in the limbic system[8]. There are many receptors in cortical regions, including the cerebral cortex - a finding which stimulated hypotheses relating anti-convulsant effects of benzodiazepines to cortical inhibition. All brain structures investigated have some BZ receptors in varying numbers, the lowest being in white-matter areas such as the corpus callosum. Even the retina contains BZ receptors, though their significance is not known.

NEW TYPES OF BZ RECEPTOR LIGANDS

There seems to exist at least three overlapping groups of (BZ) receptor ligands[9], of which one group comprises benzodiazepine like ligands, in the following referred to as agonists. Binding of this group of ligands to BZ receptors install the classical anticonvulsant, sedative and anxiolytic effects of minor tranquillizers. High affinity binding of benzodiazepine like ligands has been extensively studied. GABA agonist such as GABA and muscimol enhance the binding affinity of benzodiazepine agonist[10], in some cases the affinity for BZ receptors is about doubled by GABA agonists. A second group of BZ receptor ligands are benzodiazepine receptor antagonists; these ligands binds to BZ receptors and inhibit pharmacological and other effects elicited via the BZ receptor recognition site. Several benzodiazepine receptor antagonists have been described, for example β-CCE, PrCC, Ro 15-1788 and CGS 8216; these have all been investigated using high affinity binding techniques[11-14]. The pharmacological profile and the binding characteristics of these four receptor antagonists are similar but clearly not identical. Apparently, they all bind to the benzodiazepine recognition site, but GABA agonists fail to substantially enhance the binding affinity of the compounds[11-20]. A third group of BZ receptor ligands comprise convulsive (and/or anxiogenic) ligands, tentatively called inverse agonists. Inverse agonists binds to BZ receptors, but install exactly the opposite effects as compared to benzodiazepine agonists. To our knowledge, no other examples in pharmacology have been reported, where receptors respond in opposite

directions depending on the nature of the agonist. Two compounds of
the inverse agonist type have recently been described, methyl
β-carboline-3-carboxylate (β-CCM)[20],[21] and methyl 6,7-dimethoxy-4-
ethyl-β-carboline-3-carboxylate (DMCM)[9].

In this presentation we will discuss some of the distinct
features of the binding of DMCM and β-CCM to BZ receptors, furthermore the difference in binding properties of DMCM and β-CCM as
compared to those the benzodiazepines will be related to the difference in the pharmacology of these two groups of agents.

1. DMCM and β-CCM binds to BZ receptors

Several arguments support the idea that β-CCM and DMCM binds to
BZ receptors: β-CCM and DMCM inhibit completely specific ^{3}H-flunitra-
zepam (^{3}H-FNM) binding in several brain regions, including hippo-
campus, cortex and cerebellum[4]. Pharmacological effects of β-CCM
and of DMCM can be inhibited by benzodiazepine receptor antagonists[9],[22],[23]. The affinities for brain membranes of ^{3}H-β-CCM[20]
(K_D = 0.8 nM) and of ^{3}H-DMCM (K_D = 0.5-6 nM)[24] are in accordance
with IC_{50} values for β-CCM and DMCM, respectively, as inhibitors of
^{3}H-FNM to brain membranes. Specific binding of ^{3}H-β-CCM and ^{3}H-DMCM
have the same gross regional and subcellular distributions as
specific binding of ^{3}H-FNM[20],[24]. Several benzodiazepine receptor
active compounds inhibit specific binding of ^{3}H-β-CCM and ^{3}H-DMCM
in approximately the expected concentrations as evaluated from
inhibition of ^{3}H-FNM binding[20],[24]; conversely several compounds that
fails to inhibit ^{3}H-FNM binding also fail to inhibit ^{3}H-β-CCM and
^{3}H-DMCM binding. The total number of specific ^{3}H-β-CCM and ^{3}H-DMCM
binding sites seems to equal the number of ^{3}H-FNM binding[20],[24]. The
target size found in radiation inactivation experiments (which is
related to molecular weight) of ^{3}H-PrCC and ^{3}H-DMCM binding is
equal to that of ^{3}H-FNM (ref. 25,26, E.A. Barnard unpublished). β-CCM
and DMCM do not inhibit specific binding of several other brain
receptor ligands (ref. 9 and unpublished). Specific binding of
^{3}H-β-CCM and ^{3}H-DMCM is affected by GABA agonists indicating that
the binding sites are coupled to GABA receptors (ref. 20, 24, for
details see below). Photoshifts for BZ receptor ligands are similar
regardless of whether ^{3}H-β-CCM[27], ^{3}H-DMCM (table 2) or other BZ
receptor ligands are used as radioligand (see below).

2. GABA effects

It has recently been proposed that the ability of GABA agonists
to enhance or reduce the affinity of ligands for BZ receptors would
reflect the pharmacological efficacy of BZ receptor ligands[9],[11-20].
These studies are exemplified in table 1, which shows the GABA-ratio
for several benzodiazepine receptor ligands. A GABA-ratio above
unity. which means that GABA enhances the affinity of that ligand,
is observed for benzodiazepine like ligands, e.g. flunitrazepam.

This group of ligands are defined as having positive efficacy and
are considered to be agonists to the BZ receptor. GABA-ratios of ca.
1 reflect that the compounds have grossly the same affinity in the
presence and absence of GABA, thus, these compounds lack efficacy
and are designated as receptor antagonists. A third group of ligands,
comprising the anxiogenic (but not convulsive) ligand, FG 7142[28] and
the convulsive ligands β-CCM and DMCM have GABA-ratios below unity
(table 1). These compounds have negative efficacy at the BZ receptors
and may be named inverse agonists. From the above mentioned there
seems to be a correlation between pharmacological efficacy and the
GABA-ratio. Consequently, the GABA-ratio seems predictive for the
pharmacological efficacy of new BZ receptor ligands.

Table 1. The effect of GABA receptor stimulation on the affinity of
         ligands for BZ receptors (from ref. 9).

Ligand	GABA ratio	N
Group 1		
Flunitrazepam	2.45 ± 0.2	4
Estazolam	2.43	2
Oxazepam	2.35	2
Diazepam	2.30 ± 0.3	4
Chlordiazepoxide	2.23 ± 0.1	3
Clonazepam	2.12 ± 0.2	3
CL 218872	1.98 ± 0.1	4
Lorazepam	1.75	2
Lormetazepam	1.71	2
Zopiclone	1.53 ± 0.1	3
Group 2		
Ro 15-1788	1.22 ± 0.1	3
PrCC	1.11	2
Group 3		
β-CCE	0.86 ± 0.1	4
FG 7142	0.87 ± 0.1	4
β-CCM	0.61	2
DMCM	0.46 ± 0.1	3

A GABA-ratio of 2 means that
GABA (or muscimol) enhances
the affinity of the ligand
for BZ receptors twofold
(see ref. 9).

3. Photoshift

    Recently it has been suggested that BZ receptor agonist and
antagonists can be distinguished in another biochemical model[27].
In this model BZ receptors are inactivated by exposure to ultra-
violet light in the presence of flunitrazepam (photoaffinity
labelling). However, only 25% of the total number of benzodiazepine
sites are labelled. A major part of the remaining receptors (non-
labelled) lose their high affinity for benzodiazepines (see fig. 2),
whereas the affinity for some receptor antagonists is only slightly
affected or unaltered.

Thus, photoshifts for benzodiazepines are 0.02–0.04 (table 2) corresponding to 20–50 fold less affinity for UV/FNM exposed BZ receptors than for untreated receptors, whereas BZ receptor antagonists have photoshifts close to unity (table 2). These findings confirm results in several recent reports[27,29-32].

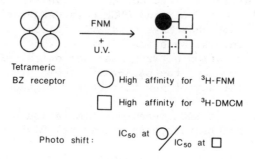

Fig. 2. Scematic representation of the postulated conformational shift in BZ receptors upon UV/FNM exposure.

Table 2. Photoshift[a] for several BZ receptor ligands using ^{3}H-DMCM as radioligand

Group	Compound	Photoshift
1, Agonist	Midazolam	0.02
	Flunitrazepam	0.03
	Lorazepam	0.03
	Triazolam	0.05
2, Receptor Antagonist	Ro 15-1788	0.85
	CGS 8216	0.93
3, Inverse Agonist	FG 7142	1.0
	β-CCM	1.4
	DMCM	2.6

a. The photolabelling procedure was done by pre incubation of washed rat cortex membranes with FNM (10 nM, final conc.) for 15 min followed by exposure to long wave UV light for 10 min and subsequent washing as described in detail[24]. The assays were done in triplicate using ^{3}H-DMCM (0.1 nM) as a radioligand as described[24]. IC_{50} values in photolabelled and normal membranes were obtained from a Hill analysis of the inhibitory effect of at least three different concentrations of the test substances. The photoshift is calculated as the IC_{50} value before UV/FNM exposure divided by the IC_{50} value obtained after treatment.

From the above conjectures it is anticipated that DMCM would have a particular high photoshift; this was confirmed because DMCM exhibited the highest photoshift yet measured (table 2). There exist, however, a few BZ receptor ligands, which do not fit well into this scheme. Zopiclone, for example, has a intermediate value for the photoshift albeit this compound possess pharmacological properties almost indistinguishable from those of the benzodiaze-pines[27]. Extended series of BZ receptor ligands, ideally including several chemical types should be investigated before the photoshift can be regarded as a reliable predictor of pharmacological efficacy.

4.   Chloride channel related interactions

The affinity of BZ receptors for ^{3}H-diazepam is increased in the presence of anions that penetrate chloride ionophores[33,34]. Likewise, some barbiturates and other agents believed to interact directly with chloride channels, enhance binding of ^{3}H-diazepam and ^{3}H-flunitrazepam to BZ-receptors (for review ref. 35). These findings point to a link between BZ receptors and the chloride ionophor related to GABA receptors. If the increase in binding in the presence of chloride ions reflects the pharmacological efficacy of BZ receptor ligands, we would expect that chloride ions would reduce binding affinity of inverse agonists, e.g. DMCM. This was not experimentally verified. Chloride ions, on the contrary, enhanced specific binding of ^{3}H-DMCM to hippocampal membranes (unpublished).

DISCUSSION

It is generally accepted that benzodiazepines exert the majority of their pharmacological effects via the GABA-system[36]; the mole-cular basis of this interaction is probably the GABA/BZ-receptor-chloride channel complex (fig. 3).

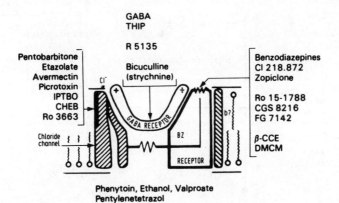

Fig. 3. A scematic representation of the GABA/BZ receptor chloride
        channel complex. Shown are drugs presumed to interact with
        the complex (from ref. 37).

In this complex the BZ receptor might be functionally similar
to allosteric sites in enzymes, which are sites distinct from the
catalytic site (which could be likened with the GABA receptor/
chloride channel). The compound that binds to the allosteric site is
named the effector. At constant enzyme and substrate concentrations,
the binding of negative effectors reduce the reaction rate of
enzymes (allosteric inhibition); the binding of positive effectors
increase the reaction rate (allosteric activation). If the allosteric effector is the substrate molecule itself, the effector is
said to be homotropic, if the effector is other than the substrate,
the effector is said to by heterotropic. Thus, by analogy to enzyme
biochemistry, benzodiazepines correspond to positive heterotropic
effectors; they enhance GABA neurotransmission indirectly. Likewise,
DMCM and β-CCM would correspond to negative heterotropic effectors.
The concept of allosteric mechanism has previously been applied to
pharmacological systems. For example the inhibitory effect of toxins
on cholinergic transmission at the nicotine receptor and also of
local anesthetics have been described as the result of negative
heterotropic effectors[38]. However, there were no indications for the
existence also of positive heterotropic effectors, which acted on
the same allosteric site. To our knowledge BZ receptor is the first
pharmacological example known, where both positive and negative
heterotropic effectors exist. The pharmacological term for a positive
effector would be an agonist (agonist = drug that produce a
response[39]). However, also the agents with negative efficacy
(negative effectors, i.e. DMCM) produce a response and are therefore
agonists, the term inverse agonist would signify that the reponse is
opposite to that of already known agonists. Partial agonists and
partial inverse agonists fit nicely into this scheme. Receptor
antagonists inhibit reponse both to partial and full agonists of the
normal and the inverse type by occupying the receptors.

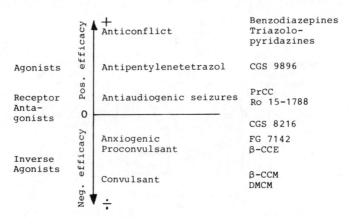

Fig. 4. A continuum of BZ receptor ligands in relation to their
        efficacy (agonists, receptor antagonists and inverse
        agonists).

An alternative to the existense of inverse agonists would be
the presence of a tonically active endogenous substance which
interact with the BZ receptor. In that case the above mentioned
receptor antagonist might represent partial agonists and the inverse
agonists (or the agonists!) might represent the true antagonists to
the endogenous ligand and thereby produce a response without having
any efficacy. However, several findings support the idea that the
present findings can be explained by the existence of agonist,
inverse agonist and receptor antagonist. First, the GABA-ratio is
unity for receptor antagonists, that means, the washed membrane
receptor preparation equilibrates in the middle position corresponding to agents having no efficacy, and not in one of the extremes,
which would be expected if a soluble endogenous ligand was removed.
Second, in well washed cultured neurones, where no tonic release of
endogenous ligand is likely to occur, the GABA response is enhanced
by benzodiazepine agonists and reduced by inverse agonist[40-44]
This again suggest that the equilibrium without endogenous ligand is
in the middle position. A spectrum of BZ receptor ligands with
partial and full positive and negative efficacies is exemplified in
fig. 4. Note that PrCC and Ro 15-1788 have weak positive efficacy
whereas CGS 8216, FG 7142 and β-CCE have weak negative efficacy.

There has been some uncertainty as to how the GABA-ratio relates
to pharmacological efficacy in terms of mechanisms. The biochemical
observation is that GABA enhances the affinity of benzodiazepine
agonist, while the fact to be explained is that benzodiazepine
agonists enhance the effects of GABA. Benzodiazepine agonists do
enhances GABA receptor affinity in a receptor antagonist sensitive
fashion, but only slightly and only in certain conditions[45-48].
These two set of results can be explained adapting the two state
concept of receptor function[9,49], which is visualized in fig. 5.

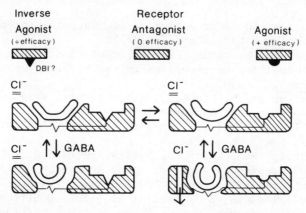

Fig. 5. A two state model for BZ receptors, relative to GABA-
        receptors and chloride channels (this model reflects
        fig. 5 in ref. 4; see Discussion for further details).

The BZ receptor is assumed to occur in two forms, the active form
(fig. 5, right) and the inactive form (fig. 5, left), these forms
are in equilibrium under normal conditions, with both forms being
present in appreciable amounts even without an endogenous or
exogenous ligands being present at the BZ recognition site. The
presence of benzodiazepine agonist, which have high affinity for the
active form will change the equilibrium to the right, thereby
increasing the percentage of the activated state necessary for the
GABA/chloride channel to be operative in GABAergic neuro-trans-
mission; GABA itself will also shift the equilibrium to the right.
BZ receptor antagonists will not affect the equilibrium, because
they have similar affinity to both forms of the BZ receptor and
consequently they will not change GABAergic transmission. Inverse
agonists have the opposite selectivity as compared to benzodiazepines and will shift the equilibrium to the left reducing GABAergic
transmission. Note that agonists by themselves may not open or close
chloride channels. This scheme (figs. 3 and 5) is consistent with
the observed enhancement and reduction of GABA effects in cultured
neurons by agonists and inverse agonists, respectively[40-44].

It is not known at present how a tetrameric BZ receptor complex
might be incorporated into a GABA/BZ receptor chloride channel
complex. Furthermore, there is at present no simple way to incorporate the results found in photoshift experiments (table 2) into a
theoretical model that describes how the photoshift might relate to
pharmacological efficacy.

Acknowledgement: We thank L.H. Jensen and E. Petersen for giving us
                 access to unpublished findings showing inhibition
                 of audiogenic seizures by Ro 15-1788.

REFERENCES

1.   R.F. Squires and C. Braestrup, Benzodiazepine receptors in
     rat brain, Nature 266:732 (1977).
2.   H. Moehler and T. Okada, Properties of ^{3}H-diazepam binding
     to benzodiazepine receptors in rat cerebral cortex,
     Life Sci. 20:2101 (1977).
3.   C. Braestrup and M. Nielsen, Benzodiazepine receptors, in:
     Handbook of Psychopharmacology, vol. 17, L.L. Iversen,
     S.D. Iversen, and S.H. Snyder, eds. pp. 285-384. Plenum
     Press, New York (1982).
4.   C. Braestrup, R. Schmiechen, M. Nielsen, and E.N. Petersen,
     Benzodiazepine ligands, receptor occupancy, pharmacological
     effect and GABA receptor coupling, in: Pharmacology of
     Benzodiazepines, S.M. Paul et al., eds. McMillan
     Press (in press).

5. R.F. Squires, D.I. Benson, C. Braestrup, J. Coupet, C.A.
    Klepner, V. Myers, and B. Beer, Some properties of brain
    specific benzodiazepine receptors: New evidence for
    multiple receptors. Pharmacol. Biochem. Behav. 10:825 (1979).

6. C. Braestrup and M. Nielsen, Multiple benzodiazepine receptors,
    Trends Neurosci. 3:301 (1980).

7. W. Sieghart and M. Karobath, Molecular heterogeneity of benzo-
    diazepine receptors, Nature (London) 286:285 (1980).

8. W.S. Young, M.J. Kuhar, Autoradiographic localization of benzo-
    diazepine receptor in the brains of humans and animals.
    Nature 280:393 (1979).

9. C. Braestrup, R. Schmiechen, G. Neff, M. Nielsen, and E.N.
    Petersen, Interaction of convulsive ligands with benzo-
    diazepine receptors, Science 216:1241 (1982).

10. J.F. Tallman, J.W. Thomas, and D.W. Gallager, GABAergic
    modulation of benzodiazepine binding site sensitivity,
    Nature 274:383 (1978).

11. M. Nielsen, H. Schou, and C. Braestrup, ^{3}H-Propyl β-carboline-
    -3-carboxylate binds specifically to brain benzodiazepine
    receptors, J. Neurochem. 36:276 (1981).

12. J. Patel, P. Marangos, and F. Goodwin, [^{3}H]Ethyl-β-carboline-
    3-carboxylate binding to the benzodiazepine receptor is
    not affected by GABA, Eur. J. Pharmacol. 72:419 (1981).

13. H. Möhler and J.G. Richards, Agonist and antagonist benzo-
    diazepine receptor interaction in vitro, Nature 294:763
    (1981).

14. A.J. Czernik, B. Petrack, H.J. Kalinsky, S. Psychoyos, W.D.
    Cash, C. Tsai, R.K. Rinehart, F.R. Granat, R.A. Lovell,
    D.E. Brundish, and R. Wade, CGS 8216: Receptor binding
    characteristics of a potent benzodiazepine antagonist.
    Life Sci. 30:363 (1982).

15. F.J. Ehlert, W.R. Roeske, C. Braestrup, S.H. Yamamura, and
    H.I. Yamamura, γ-Aminobutyric acid regulation of the
    benzodiazepine receptor: Biochemical evidence for pharma-
    cologically different effects of benzodiazepines and propyl
    β-carboline-3-carboxylate, Eur. J. Pharmacol. 70:593 (1981).

16. M. Fujimoto, K. Hirai, and T. Okabayashi, Comparison of the
    effects of GABA and chloride ion on the affinities of
    ligands for the benzodiazepine receptor, Life Sci. 30:51
    (1982).

17. A. Doble, I.L. Martin, and D.A. Richards, GABA modulation
    predicts biological activity of ligands for the benzodia-
    zepine receptor, Brit. J. Pharmacol. 76:238P (1982).

18. P. Skolnick, M.M. Schweri, E.F. Williams, V.Y. Moncada, and
    S.M. Paul, An in vitro binding assay which differentiates
    benzodiazepine "agonists" and "antagonists", Eur. J.
    Pharmacol. 78:133 (1982).

19. J.H. Skerritt, L.P. Davies, S. Chen Chow, and G.A.R. Johnston, Contrasting regulation by GABA of the displacement of benzodiazepine antagonist binding by benzodiazepine agonists and purines, Neurosci. Lett. 32:169 (1982).

20. C. Braestrup and M. Nielsen, GABA reduces binding of ^{3}H-methyl β-carboline-3-carboxylate to brain benzodiazepine receptors. Nature 294:472 (1981).

21. B.J. Jones and N.R. Oakley, The convulsant properties of methyl β-carboline-3-carboxylate in the mouse. Brit. J. Pharmacol. 74:884P (1981).

22. M. Schweri, M. Cain, J. Cook, S. Paul, and P. Skolnick, Blockade of 3-carbomethoxy-β-carboline induced seizures by diazepam and the benzodiazepine antagonists, Ro 15-1788 and CGS 8216, Pharmacol. Biochem. Behav. 17:457 (1982).

23. A. Valin, R.H. Dodd, D.R. Liston, P. Potier, and J. Rossier, Methyl-β-carboline-induced convulsions are antagonized by Ro 15-1788 and by propyl-β-carboline, Eur. J. Pharmacol. 85:93 (1982).

24. C. Braestrup, M. Nielsen, and T. Honoré, Binding of [^{3}H]-DMCM, a convulsive benzodiazepine ligand, to rat brain membranes. Preliminary studies. J. Neurochem. (1982) (submitted).

25. L.R. Chang, E.A. Barnard, M.M.S. Lo, and J.O. Dolly, Molecular sizes of benzodiazepine receptors and the interacting GABA receptors in the membrane are identical, FEBS Lett. 126:309 (1981).

26. M. Nielsen, T. Honoré, and C. Braestrup, Enhanced binding of the convulsive ligand, DMCM, to high-energy irradiated benzodiazepine receptors; evidence of complex receptor structure, Biochem. Pharmacol. (1982) (in press).

27. M. Karobath and P. Supavilai, Distinction of benzodiazepine agonists from antagonists by photoaffinity labelling of benzodiazepine receptors in vitro, Neurosci. Lett. 31:65 (1982).

28. R. Dorow, β-carboline monomethylamide causes anxiety in man. 13th Collegium Internationale Neuro-psychopharmacologicum Congress (Jerusalem, June 20-25, 1982): Abstr. I.

29. H. Möhler, Benzodiazepine receptors: Differential interaction of benzodiazepine agonists and antagonists after photo-affinity labelling with flunitrazepam. Eur. J. Pharmacol. 80:435 (1982).

30. R.W. Gee and H.I. Yamamura, Differentiation of [^{3}H] benzodia-zepine antagonist binding following the photolabelling of benzodiazepine receptors. Eur. J. Pharmacol. 82:239 (1982).

31. C.L. Brown and I.L. Martin, Photoaffinity labelling of the benzodiazepine receptor does not occlude the βCCE binding site, Brit. J. Pharmacol. 77:312P (1982).

32. J.W. Thomas and J.F. Tallman, Conversion of high-affinity benzodiazepine binding sites to low-affinity by photo-labelling, Soc. Neurosci. 402 (1982) (Abstract).

33. T. Costa, D. Rodbard, and C.B. Pert, Is the benzodiazepine
     receptor coupled to a chloride anion channel? Nature
     (London), 277:315 (1979).
34. I.L. Martin and J.M. Candy, Facilitation of specific benzodia-
     zepine binding in rat brain membrane fragments by a number
     of anions, Neuropharmacology 19:175 (1980).
35. R.W. Olsen, Drug interactions at the GABA receptor-ionophore
     complex, Ann. Rev. Pharmacol. Toxicol. 22:245 (1982).
36. W. Haefely, P. Polc, R. Schaffner, H.H. Keller, L. Pieri, and
     H. Moehler, Facilitation of GABAergic transmission by drugs,
     in: "Gaba-Neurotransmitters", P. Krogsgaard-Larsen, J.
     Scheel-Krueger, and H. Kofod, eds., pp. 357-375, Munksgaard,
     Copenhagen and Academic Press, New York (1979).
37. C. Braestrup and M. Nielsen, Anxiety, Lancet II:1030 (1982).
38. T. Heidmann and J-P. Changeux, Fast kinetic studies on the
     allosteric interactions between acetylcholine receptor and
     local anesthetic binding sites, Eur. J. Biochem. 94:281
     (1979).
39. W.C. Bowman and M.J. Rand, Textbook of Pharmacology, 2nd edition,
     Blackwell Scientific Publications, Oxford (1980).
40. R.E. Study and J.L. Barker, Diazepam and (-)-pentobarbital:
     Fluctuation analysis reveals different mechanisms for
     potentiation of γ-aminobutyric acid responses in cultured
     central neurons, Proc. Natl. Acad. Sci. 78:7180 (1981).
41. R.A. O'Brien, W. Schlosser, N.M. Spirt, S. Franco, W.D. Horst,
     P. Polc, and E.P. Bonetti, Antagonism of benzodiazepine
     receptors by beta carbolines, Life Sci. 29:75 (1981).
42. D.W. Choi, D.H. Farb, and G.D. Fischbach, Chlordiazepoxide
     selectively augments GABA action in spinal cord cell
     cultures. Nature (London) 269:342 (1977).
43. C.Y. Chan, T.T. Gibbs, and D.H. Farb, Action of beta-carboline
     in flunitrazepam-photolinked cultures, Soc. Neurosci. 572
     (1982) (Abstract).
44. M. Skovgaard Jensen and J.D.C. Lambert, The interaction of
     DMCM (a β-carboline derivative) with inhibitory amino acid
     responses on cultured mouse neurones (submitted).
45. A. Guidotti, G. Toffano, and E. Costa, An endogenous protein
     modulates the affinity of GABA and benzodiazepine receptors
     in rat brain, Nature 275:553 (1978).
46. J.H. Skerritt, M. Willow, and G.A.R. Johnston, Diazepam
     enhancement of low affinity GABA binding to rat brain
     membranes, Neurosci. Lett. 29:63 (1982).
47. B.A. Meiners and A.I. Salama, A criterion for distinguishing
     benzodiazepines from their antagonists, Soc. Neurosci. 402
     (1982) (Abstract).
48. J.H. Skerritt, G.A.R. Johnston, and C. Braestrup, Modulation of
     GABA binding to rat brain membranes by alkyl β-carboline-3-
     -carboxylate esters, Eur. J. Pharmacol. (1982) (in press).
49. D.J. Nutt and P.J. Cowen, Unusual interactions of benzodiaze-
     pine receptor antagonists, Nature 295:436 (1982).

SYNAPSES WITH MULTIPLE CHEMICAL SIGNALS: NEW MODELS AND PERSPECTIVES IN DRUG DEVELOPMENT

Erminio Costa

Laboratory of Preclinical Pharmacology, N.I.M.H.

Saint Elizabeths Hospital, Washington, D.C. 20032

1. Introduction

We have known for several decades that the protracted use of drugs that prevent the activation of neurotransmitter receptors by their specific endogenous ligands is associated with a number of untoward consequences (Paton, 1954). Nevertheless, despite this awareness, drugs that block various types of transmitter receptors have been used therapeutically, in succession. As an example, Table 1 lists some such blockers that at various times have had clinical popularity but, as Table 1 shows, the final verdict on the clinical utility of the drug often has been negative on account of overwhelmingly potent side effects. One might ask what is the reason then for persevering in such an approach? The tenet that has driven neuropsychopharmacological research in the last 30 years is that drugs that affect CNS function act at synapses, where they modify the mechanisms whereby various neurons generate, transmit, and receive specific chemical signals which are used to convey specific information from one neuron to the other. Until a few years ago, it was believed that each neuron uses one neurochemical transmitter to communicate with contiguous neurons (Eccles et al., 1956), that is, communication between neurons is by code words not by a language.

Recently, histochemical studies and neurochemical analysis has demonstrated the coexistence of two or more putative neuromodulators in the same neuron, giving substance to the belief that a language could be used in neuronal communication (Kerkut et al., 1967; Brownstein et al., 1974; Cottrel, 1977, Hokfelt et al., 1980). As indicated in Table 1, a more accurate scrutiny shows that those CNS drugs which cause less damage when used on a long term basis

127

Table 1

EVALUATION OF PRIMARY TRANSMITTER AND REGULATORY RECEPTOR BLOCKERS AS THERAPEUTIC AGENTS

Drug	Site of Action	Clinical Applications	Satisfactory in Therapy	Why?
Neuromuscular transmission Blockers	Nicotinic receptors	Surgery	Yes	Mechanical Respirators
Ganglion Blockers	Nicotinic Receptors	Hypertension	No	Side Effects
β-Adrenergic Blockers	β-Adrenergic Receptors	Hypertension	Yes	Are β-receptors primary transmitter receptors?
Muscarinic Blockers	Muscarinic Receptors	Gastric Ulcer	No	Side Effects
H_1 Receptor Blockers	H_1 Receptors	Allergy	No	Side Effects
Neuroleptics	DA_1, DA_2 Receptors	Schizophrenia	No	Yatrogenic
Anxiolytics	GABA Receptors	Anxiety	Yes	Acts on a regulatory site for GABA receptor
Typical Antidepressants	5HT Uptake	Endogenous Depression	Yes	Acts on a regulatory site for 5HT uptake
Atypical Antidepressants	$5HT_2$ Receptors	Endogenous Depression	Yes	Acts on a regulatory site for $5HT_1$ receptors

show that they are working on synapses which transact synaptic
function by using more than one signal. In general, the sites
where the most successful drugs appear to act are synaptic sites
that have an accessory role in communication transaction. These
sites when activated by endogenous effectors amplify or reduce the
response elicited by those modulators that can activate the transducer system of the synapse. An explanation for this peculiar behavior is the possibility that when these sites are activated by
endogenous effectors they cannot activate the transducer but just
change the gain at which the transducer operates. The effectors of
these peculiar synaptic sites have been termed cotransmitters.

The finding that the mode in which some neurons instruct the
next neuron is not through a code word as postulated by the so-
called Dale's Law* but through at least two different chemical signals, appears to have important consequences in pharmacology,because
it could direct the conception and design for new drug development.
In this presentation, I will attempt to project into the future and
predict possible avenues that we might follow to have better and
safer drugs to help mental patients cope with reality. I will not
dwell at any length on the problem of tardive dyskinesia and other
inconveniences caused by a long term treatment with neuroleptics
because this topic was covered by a recent extensive publication to
which I refer the reader of the present manuscript (Jeste and Wyatt,
1982). I will limit my comments to state that each physician who
has spent some time in the care of psychiatric patients could trace
a good picture of the "neuroleptic man" in a similar fashion of the
effective description made by Patton (1954) of the "hexamethonium
man." While the "hexamethonium man" could keep the hypertension
under partial control, the "neuroleptic man" controls the symptoms
of the psychosis at least partially, by accepting a number of motor
deficiencies and muscle function abnormalities. Hence, the over-
riding question is, do we have any valid alternatives to correct
a psychosis other than by changing the capacity to control motor
output of patients? My answer to this question is that we do not
have any alternatives, but we do have hope for a perspective al-

* Sir H. Dale in 1935 during the Dixon's lecture (Dale 1935) stated
that if a neuron of the sensory ganglia was releasing a given transmitter at one terminal, it would release the same transmitter at all
the others.   This statement was not meant to indicate that the
neuron was secreting only one transmitter, but when this statement
was reinterpreted by Sir John Eccles and coworkers (1956) it became
a statement of the uniqueness of neurotransmitters in any given
neuron and went to history as Dale's Law. It may be of interest to
those of you who do not believe in preconceived ideas of journal
references to know that one of the first papers showing that a neuron stores more than one neuroeffector (Brownstein et al., 1974)
was first rejected by a British journal because it contravened
Dale's Law.

ternative. Hence, I will try to give some background on which my
hopes are based, still stressing that I will not promise a therapy
for psychoses but only a remedy for its symptoms. In fact, I firmly
believe that before thinking about therapy of psychoses, we must
first reach a better understanding not only of the basic factors of
these diseases, but also we must comprehend more extensively the
basic mechanisms of brain function.

2. Receptor models for synapses which operate with single or multiple chemical signals

Synaptic transaction of neuronal communication is the result of
an interaction of two major morphological components at interneuronal contact points: a presynaptic functional unit from which specific
chemicals are emitted and a postsynaptic unit for receiving these
signals termed receptor, which is located in the membrane of the
postsynaptic cell. These signals are called neurotransmitters and
they are released extraneurally to transact neuronal communication
The specific postsynaptic receptors sense the chemical signals that
are released from the presynaptic unit and transduce these signals
into specific metabolic stimuli for the postsynaptic cell (Fig.1).
Each synapse has the capability to enhance or restrict the gain at
which the communication is transacted and uses this capability to
harmonize its output to the characteristics of the input it receives.
This process is optimized through a modification of the functional
output of the presynaptic or the postsynaptic units via specialized
mechanisms, which as a complex in their functional output are referred to as "synaptic plasticity." This term indicates the capability of synapses to adapt to environmental changes by using molecular mechanisms which are not yet understood in their complexity.
We just can propose some models to study the most simplified aspects of this complex integrative function. We believe that the
presynaptic units by changing synthesis rate of transmitters contribute to those plasticity phenomena that have a time constant in the
order of minutes, while postsynaptic units may be operative in
faster plasticity changes. These fast mechanisms that are included
under the term of synaptic plasticity adapt neuronal function to
rapid changes and provide a continuum of functional plasticity.
Perhaps, these rapid changes are possible by harmonizing the functional subunits of the receptor with specific chemical signals.
Some signals become operative by changing the properties of the
transmitter recognition sites; others do so by harmonizing receptor coupling devices and still others by changing the metabolism
of membrane phospholipid. The latter changes are often referred
to as changes in the receptor microenvironment: they are very important. Because of the supramolecular organization of the receptors, their functional subunits operate within the lipid bilayer of the postsynaptic membrane and their capability of interacting with each other can be either maximized or minimized "inter
alia" by changes in lipid fluidity. As mentioned above, the trans-

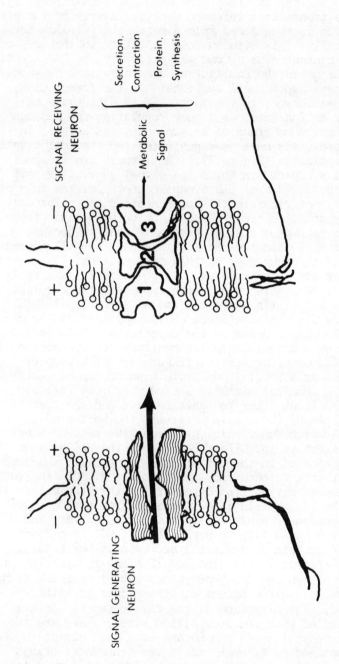

Figure 1

Schematic representation of intercellular communication using a single chemical signal -

1= recognition site for the transmitter; 2= coupler mechanism; 3= transducer which transforms the chemical signal generated by the presynaptic site into a metabolic signal for the post-synaptic cell. The coupler mechanism is not an essential feature of receptors and 1 and 3 at times can be even located in the same protein.

ducer is the unit of the postsynaptic receptor complex that is responsible for changing the nature of the chemical signal (Fig. 1). This signal is the transmitter released extracellularly by the presynaptic unit, which is transduced into a metabolic stimulus for the postsynaptic cell. This stimulus can be either an increase of the intracellular content of a second messenger (cyclic AMP, cyclic GMP, etc.) or a change in the equilibrium concentrations of specific ions across the postsynaptic cell membrane. In the former case, the transducer is an enzyme, in the latter it is either a specific ion carrier or an ion gate, or a pump regulating the passage across the postsynaptic membrane of an ion, or of an amine. In conventional synapses, the unit that generates the chemical signal is located presynaptically (Figure 1). But, there are synapses called "reciprocal synapses" in which the signal generating unit can be located on both sides of the synaptic gap. For the sake of simplicity, we will not refer in this presentation to reciprocal synapses. In a synapse, multiple signals can occur because there are multiple units each producing a chemical signal or because one unit generates more than one chemical signal. We will address such a situation in the remainder of this presentation. The chemical signal whether single or multiple, is released extraneurally by a depolarizing wave generated by nerve impulses; the chemical signal(s) bind to a specific detector, and as a result of such binding, it changes the conformational characteristics of the detector. These molecular changes in the detector are transferred to the transducer by a translocation mechanism which operates by virtue of a specific coupling device. (Figure 1). The coupler mechanism functions as a specialized protein which can be activated by a specific chemical signal. The coupler can relate equally well to the transmitter recognition site and the transducer (Figure 1). Among the coupler systems, the best studied is the G/F protein(s) which function(s) by having two sets of binding sites which are selective for GTP or GDP. When the affinity of the GDP binding sites located on the G/F protein is maximal, the coupler associates with the detector for the transmitter. During such an association the affinity of the detector for the transmitter is lower than when the G/F protein is dissociated from the detector. However, when the detector binds the specific transmitter, the G/F protein acquires the capability to bind GTP. If GTP is present its binding to the G/F protein determines a decrease in GDP binding affinity and the G/F protein translocates from the recognition site to the transducer. When the G/F protein associates with the transducer its response to a given amount of transmitter is either reduced or amplified. The direction of the change depends on the type of G/F protein coupled with the recognition site. There are two types of G/F binding protein (I and S) and both can coexist in the same receptor; the binding of each one to the transducer brings a different message: a) one which operationally we will call inhibitory (I), where the G/F protein deemphasizes the gain of the

transducer's operation; and b) the other called stimulatory (S)
where the G/F protein emphasizes the gain of the transducer's operation. In synapses with multiple signals, the two proteins can be
located in the domain of the same transducer but each one is activated by a specific chemical signal. Since the two proteins have
access to the same transducer the two chemical signals can modulate
the transducer activity in opposite directions (Fig. 2). Hence,
when multiple signals are generated, the resulting response depends
on the ratio of the quantity of the two signals that can activate
or reduce the gain of the transducer operation. The molecular mechanisms involved in this operation are not known: some data support
the view that the extent at which the transducer function is modulated depends on the activity of the specific GTPase (possibly located in the G/F protein) which can regulate the concentration of
the two specific allosteric activators of the G/F protein. Many
examples are available in which the G/F protein(s) are the coupler
mechanisms for transmitter receptors that have as a transducer,
adenylate cyclase, the enzyme that generates cAMP. Transmitters
that modulate the operation of specific carriers, gates, or pumps,
may have another type or types of couplers. We are currently
studying a membrane protein, GABA-modulin, which is probably functionally associated with GABA receptors and interacts with GABA
recognition sites (Guidotti et al. (1982), reducing the number of
high affinity GABA recognition sites that are available (Fig.3).
GABA-modulin is phosphorylated by protein kinases and appears to
be coupled with these enzymes functionally (Wise et al., 1983).
There are five sites in GABA-modulin that can be phosphorylated by
the protein kinases: the Ca^{2+} dependent enzyme phosphorylates only
one of these sites whereas cyclic AMP dependent protein kinase
phosphorylates the other four sites. When GABA-modulin is phosphorylated by Ca^{2+} dependent protein kinase, the GABA-modulin can
still suppress the number of high affinity GABA recognition sites
operative in the postsynaptic membrane. When GABA-modulin is phosphorylated by the cAMP dependent protein kinase, four different
other sites of the protein are phosphorylated and the protein as-
sumes a conformation that opposes its association with the process
that reduces the number of GABA recognition sites with high affinity. It can provisionally be inferred that when cAMP dependent
phosphorylation occurs, high affinity GABA recognition sites can
associate with the transducer thereby emphasizing the gain of the
process whereby GABA opens the Cl- gate. Perhaps, a Ca^{2+} dependent
protein kinase activated by a voltage dependent Ca^{2+} channel, phosphorylates GABA-modulin allowing its association with the high
affinity GABA recognition site. Conversely, when the cAMP-depend-
ent protein kinase is activated as a result of the action of a cotransmitter, GABA recognition sites with high affinity become available for the action of GABA because GABA-modulin cannot reduce
their availability (Fig. 3).

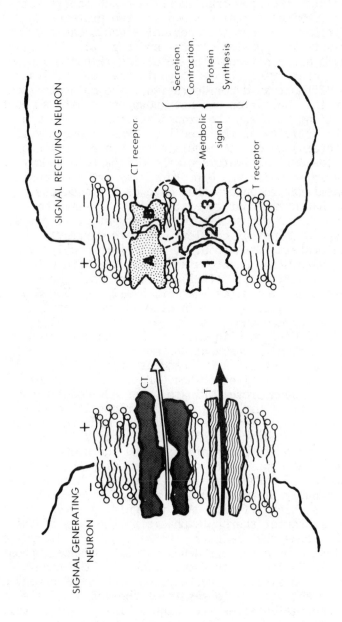

Figure 2

Schematic representation of intercellular communication using a multiple chemical signal -

T receptor = receptor for transmitter (T) which is the signal that can activate the transducer
(3). CT receptor = receptor for the cotransmitter (CT) which is the signal that can modulate
the gain of the transducer through A and B

3.   Pharmacology of synapses with multiple signals

The consequence of the coexistence of more than one modulator
represent an interesting focal point that can be considered in
the development of new drugs after some of the current perplexities
are clarified. First of all, we must recognize that the presence
of a neuromodulator in a presynaptic site does not per se mean that
this compound functions as a neurotransmitter. Its release and
the presence of a specific synaptic receptor must be documented
before considering the participation of multiple signals in the
synaptic transaction. Once we are sure that a synapse functions
with more than one signal, then there is the possibility that the
signals operate independently. If they interact, the cotransmitter
functions by: 1) amplifying or reducing the gain of the synaptic
action of the transmitter; or 2) acting on a process that modifies
the gain of the synaptic action of the transmitter; or 3) acting

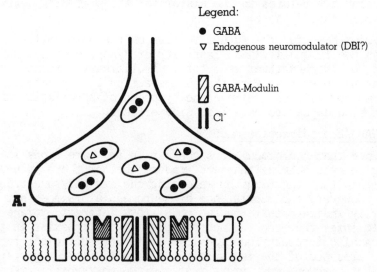

Legend:

● GABA

▽ Endogenous neuromodulator (DBI?)

GABA-Modulin

Cl⁻

Figure 3: A model depicting GABA terminal with primary transmitter
and endogenous modulator (DBI). Postsynaptically, there is a DBI
and a GABA$_A$ recognition site each one modeled after the signal they
detect. We know that phosphorylation of GABA modulin prevents the
interaction of GABA recognition sites with the Cl- but we ignore
the molecular mechanism whereby GABA modulin DBI and GABA recognition sites interact with each other. Hence, the arrangement of
these units in the cartoon does not depict a possible location in
the membrane.

on a process that modifies the lipid fluidity of the receptor domain; thereby modifying indirectly receptor function; or 4) changing in the opposite direction the response of the transducer to the
transmitter (Fig.2). The gain at which the transducer operates can
be set by the operation of the coupler system which receives in-
structions from the recognition site of the primary transmitter.
The informational structure of the presynaptic chemical signal can
be of two types: the primary information is that it is necessary
for the activation of the transducer; however, not all the signals
that are generated presynaptically have primary informational structures. Some of them instruct the same receptor system, giving gain,
duration, or direction to the response of the primary transmitter.
Hence, contrary to the traditional belief that the information process generated by the presynaptic neuron arrives as a set of signals with specified modalities, the coexistence of two modulators
creates the possibility that the primary transmitter input is specified but the associated cotransmitter input is elaborated in a variable way by modalities that we do not entirely understand, and
"inter alia" are related to the history of the preceding activity
operative in that synapse.

The great hope for the future resides on the possibility that
drugs can modify quality and plasticity of synaptic transmission
by acting on cotransmitter receptors. Preliminary evidence is en-
couraging because it appears that drugs acting on cotransmitter
mechanisms not only can relieve the symptoms of psychiatric disorders but also appear to have less side effects.

A. The GABA/benzodiazepine model

Evidence that neuromodulators coexist in the same axons is now
becoming so overwhelming that one wonders whether there is any neuron in brain that functions by releasing one and only one transmitter. However, the coexistence of neuromodulators does not
necessarily indicate that the two modulators interact extraneuron-
ally. To infer that the two neuromodulators that coexist in the
same axon also interact functionally, it is not sufficient to show
only the presence of the two specific recognition sites postsynaptically but also one must prove that the two postsynaptic receptors
are functionally linked as depicted in the cartoon of Fig.2. This
evidence should be histochemical, pharmacological, and electrophysiological. Histochemically, by using photoaffinity labeling
with radioactive receptor ligands, one can definitely show that
two neuroactive substances coexisting in the presynaptic unit have
contiguous postsynaptic receptors. Though this histochemical
evidence is always a necessary prerequisite, it is not per se
sufficient to infer that the two contiguous receptors interact
functionally. A functional receptor interaction can be determined
by establishing a reciprocal or unilateral modification of the receptor characteristics following the occupancy of one of the two re-

ceptors by specific ligands, and by measuring the characteristics of the ion channels following agonist occupancy of the receptors.

A number of studies have been published inferring that at least in a subpopulation of GABA receptors, this primary transmitter function can be modulated by multiple chemical signals (Costa and Guidotti, 1979). This consideration was prompted by accumulating evidence of GABA's participation in the anxiolytic, anticonvulsant and muscle relaxant actions of benzodiazepines (Costa et al., 1975). Since direct studies had shown that benzodiazepines do not bind to GABA recognition sites and do not effect the synthesis, release and reuptake of GABA (Haefely et al., 1979) it became necessary to consider whether benzodiazepines modify GABA recognition sites. Direct experiments indicated that the Bmax of the high affinity recognition sites for GABA can be increased by preincubation of the membranes with μM concentrations of benzodiazepines (Costa et al. 1978; Guidotti et al., 1978; Baraldi et al., 1979). Since the hgih affinity recognition sites for GABA are important to activate the GABA dependent Cl- channels (Guidotti et al., 1979) this benzodiazepine modification of GABA recognition sites could have functional implication. Such an implication became more plausible when Mohler and Okada (1978) and Braestrup and Squires (1978) reported that benzodiazepines bind with high affinity to crude synaptic membranes prepared from the brain of several species including man and that in a series of benzodiazepines this affinity for this binding site was proportional to their therapeutic potency. Thus, not only this high affinity binding has pharmacological relevance in the anxiolytic action (Haefely et al., 1981) but also these sites may have a role in the action of some endogenous effector that regulates GABA receptor function, physiologically. The information available ,taken together strongly suggest that the recognition site for the benzodiazepines could be a site for the action of a cotransmitter modulating GABA synapses.

Starting from this inference, we have isolated from brain a polypeptide that has been purified to homogeneity (Table 2) and has a set of important pharmacological and biochemical properties listed in Table 3. At this time, it is possible to say that in brain there is an endogenous effector of benzodiazepine recognition sites which acts as an antagonist of the anxiolytic action of benzodiazepines and which is perhaps anxiogenic (Table 3). Keeping in mind that anxiety building could have a survival value, it is possible to suggest that anxiety can be modulated by the endogenous peptide we are studying or perhaps by some other peptide (either a metabolite of this peptide or one unidentified yet). We are now preparing antibodies to this endogenous anxiogenic effector of the benzodiazepine recognition sites to ascertain whether it is present in human brain and spinal fluid. Should it be present in cerebral spinal fluid then one could study whether its content

Table 2

PURIFICATION FROM BRAIN OF AN ENDOGENOUS EFFECTOR FOR THE HIGH
AFFINITY BINDING SITES FOR BENZODIAZEPINES LOCATED IN CRUDE
SYNAPTIC MEMBRANES

1. Homogenization in 20 volumes of 1 N acetic acid 90° for 10 min.
2. 48,000 x g supernatant adjusted to pH 5.
3. 48,000 x g supernatant hyophilized
4. Sephadex G-100 column chromatography
5. Sephadex G-75 column chromatography
6. Precipitation of contaminant material with 60% annonium
    sulphate saturation.
7. 48,000 x supernatant on Biogel P2 column
8. HPLC on Synchropak AX 300 anion exchange column
9. HPLC reverse phase on Bio Sil ODS 10 column (acetomitrile
    gradient 0-60%).
10. HPLC reverse phase on Bio-Sil ODS 10 column (35% acetonitrile
    isonotic).
11. One band in polyacrylamide gel electrophoresis.

Table 3

CHARACTERISTICS OF ENDOGENOUS PUTATIVE LIGAND FOR DIAZEPAM BINDING

Molecular Weight	9500 Daltons
Thermostability	Yes
Resistance to tryptic digestion (2 hrs at 37°)	No
Resistance to pronase digestion (2 hrs at 37°)	No
K_1 for displacing ^{3}H-diazepam competitively	$0.5 \mu M$
Carboxy terminal free	Yes
N terminal free	No
Displaces or binds GABA	No
Binds ^{3}H-benzodiazepines	No
Dialyzable	No
Degradation by phospholipase (A,B,C)	No
Precipitated by $(NH_4)_2SO_4$	No
Has proconflict activity (i.c.v.)	Yes
Has anticonflict activity (i.c.v.)	No
Antagonizes anticonflict action of BZD (i.c.v.)	Yes

changes in mental illness; for instance, in pathological anxiety
and in individuals with character disorders.

Before giving physiological importance to the benzodiazepine
recognition sites, a series of studies were initiated to elucidate whether these binding sites were located close to GABAergic
synapses. By using histochemistry and a double staining technique
for the immunochemical detection of glutamic acid decarboxylase

combined with photoaffinity labeling of the benzodiazepine recognition site with ^{3}H-flunitrazepam, it was shown that these recognition sites were always located in conjunction with GABAergic axons (Mohler et al., 1980,81). This relationship was corroborated by the finding that the central action of benzodiazepines depends on the presence of an optimal amount of GABA in the axon terminals (Biggio et al., 1977). When the study of GABA-benzodiazepine interaction at receptor level was extended to cultures of spinal cord neurons using fluctuation analysis, it was found that benzodiazepines as expected for a cotransmitter, could not open the GABA dependent Cl channel. This transducer could be activated only by the addition of GABA (Study and Barker, 1981). However, benzodiazepines added together with GABA could increase the frequency of the Cl- channel opening elicited by GABA (Study and Barker, 1981). This finding was in keeping with the view that GABA acts as a primary transmitter capable of activating the transducer (Cl- channel) whereas benzodiazepines because they act on a cotransmitter recognition site cannot activate the transducer, but they change the gain at which GABA receptors operate. Further support to this view comes from experiments showing that in vivo GABA can facilitate the function of benzodiazepine binding sites by increasing their affinity for specific benzodiazepines (Tallman et al., 1978); conversely, the addition of benzodiazepines to crude synaptic membranes in the presence of EGTA increases the B_{max} of the high affinity GABA binding (Guidotti et al., 1978; Baraldi et al., 1979; Majewska et al., 1982).

Since the recognition sites for GABA and benzodiazepines are located in two independent peptides, the interaction evidenced by the above mentioned experiments can be seen as an intermolecular process, involving the two recognition sites and GABA modulin, the specific coupler for GABAergic synapses mentioned above. In fact, the facilitation of ^{3}H-muscimol binding to synaptic membranes elicited benzodiazepines is inhibited by GABA modulin, This inhibition reflects a specificity related to the site where benzodiazepines act, in fact, GABA modulin fails to inhibit the facilitation of ^{3}H-muscimol binding elicited by pentobarbital (Table 4).

Table 4
FACILITATION OF ^{3}H-MUSCIMOL BINDING BY BENZODIAZEPINES OR (-)-
PENTOBARBITAL: ROLE OF GABA MODULIN

Facilitating	Concentration	GABA-modulin (2.5 μM)	Facilitation (%)
Diazepam	10^{-6}	No	33
Diazepam	10^{-6}	Yes	3
(-)-Pentobarbital	10^{-4}	No	42
(-)-Pentobarbital	10^{-4}	Yes	46

Membranes were prepared as described in Guidotti et al. (1982). ^{3}H-muscimol (5 nM).

All these data taken together allow for the following generali-
zations: 1) benzodiazepines and GABA recognition sites are two distinct subunits of the GABA receptor; 2) these sites interact bio-
chemically and functionally, this interaction is regulated by allosteric changes occurring to the GABA modulin, a basic protein present in synaptic membranes; 3) the effector for these allosteric
changes appears to be a phosphorylation by a cAMP dependent protein
kinase; 4) when the benzodiazepine recognition sites are occupied
by benzodiazepines, the Cl- gate fails to open. Since the Cl- gate
is the transducer function linked to GABA recognition sites, and
since the activation of benzodiazepine recognition sites can facilitate the action of GABA to open the Cl- gate but does not open
per se the Cl- gate; hence as of now the information available indicates that GABA acts as a primary transmitter and the endogenous
effector of benzodiazepine recognition site is a cotransmitter; 5)
there are three classes of ligands that bind to this cotransmitter
site: one class is endowed with anxiolytic, anticonflict and anti-
convulsant properties, another class is proconvulsant and anxiogenic (proconflict), and the third class is almost devoid of pharmacological action (Corda et al., 1981), but still blocks anxiolytic
and anxiogenic actions elicted by occupancy of benzodiazepine recognition sites (Haefely et al.,1981; Braestrup et al.,1982). The endogenous effector of benzodiazepine receptors we have isolated
appears to belong to a category of anxiogenic derivatives which have
proconflict activity (Corda et al., 1982)

B. The chromaffin cell model: are enkephalins cotransmitters
              for acetylcholine?

The new vistas on synaptic organization with multiple chemical
signals are operative also in periphery and probably apply to the
classic cholinergic synaptic mechanisms between the splanchnic
nerve terminals and the chromaffin cells. In recent years, biochemical (Costa et al., 1979) and histochemical (Hokfelt et al.,
1980) evidence has accumulated suggesting that in splanchnic axons
and in chromaffin cells, enkephalin-like material may coexist with
acetylcholine and catecholamines, respectively. More recently,
Tang et al. (1982) and Panula et al. (1982) provided biochemical
and histochemical evidence supporting the presence of the heptapeptide Met[5]-enkephalin Arg[6] Phe[7] in the chromaffin granules and
in the axon terminal of the splanchnic nerve. It has been suggested on the basis of indirect evidence (Costa et al., 1981, 1983)
that nerve impulses corelease these peptides and acetylcholine and
these two coexisting neuroactive substances may interact functionally through specific recognition sites for acetylcholine and opiate
peptides, located on the membranes of chromaffin cells (Hokfelt
et al., 1980; Kumakura et al., 1980).

A variety of ^{3}H-ligands were used to characterize the opiate
binding sites located on the membranes of adrenal medulla cells.
In addition to ^{3}H-etorphine, ^{3}H-diprenorphin and ^{3}H-naloxone speci-

fic markers for $\mu, \kappa, \sigma, \epsilon$ receptors were used. While the K_D for these ligands was not very different (always in the n molar range), the apparent maximal density of binding sites was different ranging from 230 fmoles/mg protein for ^{3}H-etorphine to 3 fmoles/mg protein for ^{3}H-[D-Ala^2-D-Leu^5]-enkephalin. ^{3}H-dihydromorphine, ^{3}H-ethylketazocine and ^{3}H-N-allynometazocine had an apparent Bmax between 18 and 30 fmoles/mg protein whereas the corresponding values of ^{3}H-diprenorphine and ^{3}H-naloxone were 170 and 172 fmole/mg protein, respectively. It is noteworthy that the number of binding sites for ^{3}H-etorphine is 2/3 of the sum of the total opiate binding sites measured with all the other ligands. We have interpreted these data to indicate that adrenal medulla does not have classical $\mu, \kappa, \sigma, \epsilon$ receptors but a novel type of receptor that specifically binds etorphine (Saiani and Guidotti, 1983).

Table 5

IDENTITY IN RANK ORDER OF THE INHIBITORY CONTENT (K_i) FOR ^{3}H-ETORPHINE DISPLACEMENT FROM MEDULLARY SYNAPTIC MEMBRANES AND THE IC_{30} TO INHIBIT BY 30% THE RELEASE OF CATECHOLAMINES ELICITED BY ACETYLCHOLINE

Opiate	K_i	IC_{30}
Etorphine	1	1×10^{-7}
Beta-endorphine	10	2×10^{-7}
[D-Ala^2-MePhe^4(0)^5-ol]	15	2.5×10^{-7}
Met^5-enkephalin Arg^{6}Phe^7	16	6.5×10^{-7}
Levorphanol	50	7.5×10^{-7}
Ethylketazocine	100	9×10^{-5}
Morphine	300	1×10^{-4}
[D-Ala^2-D-Leu^5-Enkephalin]	400	2×10^{-4}

Catecholamine secretion was elicited by acetylcholine (5×10^{-4}M) Catecholamines were measured with HPLC technique. K_i was calculated according to Cheng and Prusoff (1973). All the drugs were tested using 4-7 concentrations in triplicate with 0.5 and 1 nM ^{3}H-etorphine. The K_is were the mean of 3 to 5 experiments. The samples contained 10^{-3}M ortho phenathroline in order to prevent the degradation of the peptides. To the K_i for etorphine (nM) was attributed the nominal value of 1. The IC_{30} for catecholamines was extrapolated from 4 doses and analyses were repeated in triplicate

Additional experiments (Table 5) were performed to characterize the receptors located in membranes of adrenal medullary cells. Interestingly Met^5-enkephalin Arg^6 Phe^7, a peptide present in high concentrations in the splanchnic nerve terminal (Panula et al., 1982) and beta-endorphin are the most potent endogenous opiates tested. The inhibition of the catecholamine release by opiate

agonists was blocked by diprenorphine and naloxone suggesting that
opiate receptors are involved in the inhibition of the ACh elicited
secretion of catecholamines from chromaffin cells (Saiani and
Guidotti, 1983). In addition, the opiate inhibition was stereo-
selective because levorphanol was two orders of magnitude more potent than dextrorphan (Saiani and Guidotti, 1983). Finally, it
appears that opiates and ACh receptor sites are functionally linked
and that a specific heptapeptide may be the physiological modulator
of ACh action. Thus, by mimicking the action of the heptapeptide
one could deemphasize the ACh function. Then, the question arises,
is the transmitter-cotransmitter relationship operative in vivo?
To test this hypothesis, we studied in the dog whether the opiate
receptor antagonist diprenorphine modifies the amount of catecholamines that are released from adrenal medulla by electrical stimulation of the splanchnic nerve (Table 6).

Table 6

RELEASE OF CATECHOLAMINES (ng/ml plasma) INTO ADRENAL BLOOD ELICITED
    BY SPLANCHNIC NERVE STIMULATION (10 V)

Drug	Hz					
	0		3		9	
	E	NE	E	NE	E	NE
Saline	107	12	174	34	618	128
Diprenorphine (0.2 mg/kg i.v.)	20	3	295	50	1128	270

Technique for blood sample collection from adrenal vein and stimulation described in Costa et al. (1981). Analysis of catecholamines by HPLC with electrochemical detection method.

The data of Table 6 show that in dog, the blockade of opiate receptors facilitates the neurally mediated release of catecholamines
from adrenal medulla. These experiments suggest that opiate receptors of chromaffin cells are functionally linked to cholinergic receptors in a transmitter (ACh)/cotransmitter (opiate) relationship.
These opiate receptors are peculiar in their selectivity for specific ligands and, therefore, cannot be classified according to traditional nomenclatures. Moreover, since the intrinsic activity of
morphine or D-Ala^{2}D-Leu^5 in preventing secretory action of ACh, is
more than 100 times less than that of Met^5-enkephalin-Arg^6-Phe^7,
the opiate receptors functioning as cotransmitters for ACh cannot
be studied by using morphine or enkephalin pentapeptides as reference drugs. Met^5-enkephalin-Arg^6-Phe^7 or its stable analogue
should be preferred. The question then is: is this heptapeptide's
specificity peculiar only for chromaffin cells or does it apply to
other tissues including CNS?

Table 7

Met^5-ENKEPHALIN AND Met^5-ENKEPHALIN-Arg^6-Phe^7-LIKE IMMUNOREACTIVITY
IN RAT TISSUE

Tissue	Met^5-Enkephalin	Met^5-Enkephalin-Arg^6-Phe^7
	(pmol/mg protein±S.E.)	
Striatum	12.5±0.85	2.5±0.17
Hypothalamus	8.0±0.93	1.6±0.16
Medulla Oblongata	2.3±0.41	0.51±0.017
Midbrain	2.2±0.39	0.42±0.021
Cortex	1.8±0.21	0.26±0.031
Cerebellum	0.60±0.096	0.15±0.0053
Hippocampus	0.75±0.093	0.11±0.014
Superior Cervical Sympathetic Ganglia	0.52±0.080	1.2±0.31
Adrenal Gland	0.23±0.034	0.50±0.045
Ileum	0.86±0.11	1.2±0.22

The data of Table 7 suggests that the heptapeptide, Met^5-enke-phalin Arg^6-Phe^7 may become a very important neuropeptide because of its presence in several brain structures, in sympathetic ganglia, adrenal and ileum. One of the possibilities that must be consider-ed is that MEAP may function as an immediate precursor of Met^5-enkephalin because it could be converted into this pentapeptide by a dipeptidyl carboxypeptidase (Yang et al., 1981). Such an enzyme is located in brain tissue and can be distinguished from enkephalin-ase. For instance, thiorphan IC_{50} to inhibit enkephalinase is $10^{-7}M$ but the IC_{50} for the dipeptidylcarboxypeptidase is greater than $10^{-5}M$; in contrast, captopril IC_{50} for the latter enzyme is $0.7 \times 10^{-8}M$ while the IC_{50} for enkephalinase is greater than $10^{-5}M$. Rats injected with 30 μg of captopril intraventricularly show no change in the striatal heptapeptide and Met^5-enkephalin content. However, when the rats were injected with captopril the analgesia elicited by acupuncture was prolonged. Preliminary results indi-cate that the heptapeptide content of striatum is increased in rats receiving electroacupuncture after intraventricular captopril. Since Met^5-enkephalin-Arg^6-Phe^7 can be released from striatal slices by depolarizing concentrations of K^+ in a Ca^{2+} dependent manner (Yang et al., 1983), it is possible that the Met^5-enkepha-lin Arg^6-Phe^7 heptapeptide acts as a neuromodulator. These data taken together make a strong case that Met^5-enkephalin Arg^{6}Phe^7 acts as a neuromodulator and suggest that dipeptidylcarboxypepti-dase may be located extraneurally and act on the released hepta-peptide. Perhaps, there are some other systems in the CNS or in periphery that recognize Met^5-enkephalin Arg^6-Phe^7 as a physiolo-gical effector different from other enkephalins; thereby supporting the view that this heptapeptide is an active opioid peptide and does not require conversion to Met[5]-enkephalin for biological activity.

4. Chemical signals to regulate serotonin uptake: a model for the action of imipramine

Imipramine became the antidepressant of election through serendipity. Only after this drug was established in the clinic as an effective remedy for depression, its capacity to antagonize the reuptake of catecholamines and serotonin was demonstrated. The evidence that imipramine which ameliorated the symptoms of depression also enhanced catecholamine and serotonin availability at synaptic receptors, gave credence to the view that a deficit in monoaminergic function was the cause of depressive illness (Schildkraut 1965, 1978). However, in establishing this relationship, attention was not given to the different time courses for the blockade of uptake and antidepressant action. For the last 15 years, based on this ill-conceived inference the symptoms of affective disorders were considered to reflect either a decreased availability of catecholamines at specific postsynaptic recognition sites or a change in the physiological balance between the function of neurons that store serotonin (5HT) and norepinephrine (NE). This thinking rested on the observation that the antidepressants available to alleviate the symptoms of depression either blocked monoamine uptake or inhibited monoamineoxidase. However, it must be stressed that the inference supporting the role of monoamines in affective disorders included an important error: while the compounds blocked amine uptake and metabolism almost instantaneously the therapeutic action always occurred after an interval of at least two weeks. Because of this time course discrepancy only recently it was excluded that the blockade of uptake or metabolism per se was the pivotal element in the therapeutic action of imipramine. Presumably, the beneficial effect on depression exerted by these drugs depends on some action which is triggered by the blockade of the uptake.

The discovery that imipramine binds with high affinity to specific recognition sites preferentially located in axons of serotonergic neurons (Raisman et al., 1979; Sette et al., 1981, Palkovits et al., 1981, Brunello et al., 1982) not only has given us an important clue concerning the site at which imipramine acts but also has provided us with a tool to study the molecular mechanisms involved in the imipramine action. Usually high affinity binding sites for drugs located on neuronal membranes denote sites where endogenous effectors are physiologically operative. Since binding studies combined with specific brain lesions have identified the ^{3}H-imipramine binding sites to be located preferentially in 5HT axons (Brunello et al., 1982; Sette et al., 1981), it became necessary to identify the function which is regulated by these binding sites. In order to answer this question, we carried out an experi-

ment using the concept of receptor down regulation elicited by prolonged treatment with a drug as a criterion to distinguish high
affinity binding sites functioning in a supramolecular receptor
unit from those functioning as an acceptor which are devoid of
function. The number of recognition sites for a primary neurotransmitter can be down regulated if the ligand given for protracted
time periods functions as an agonist; conversely, the number of
binding sites can increase if the ligand injected repeatedly functions as an antagonist. However, there are the receptors for cotransmitters that cannot be down or up regulated by a measureable
extent by injections of receptor ligands given daily for several
weeks. Hence, the criterion of down regulation cannot be an abso-
lute criterion for differentiation. However, whenever this change
in the number of receptors occurs, one can use it as a criterion to
exclude acceptors. ^{3}H-imipramine binding sites can be down regulated after injections (10 mg/kg i.p. repeated twice a day for 10
days) (Kinnier et al., 1980). Using this down regulation, we
assessed the functional role of ^{3}H-imipramine binding sites by re-
sorting to two research strategies. We attempted to determine
whether the presence of ^{3}H-imipramine binding sites located on 5HT
axons was necessary for the down regulation of NE induced amplification of adenylate cyclase which is elicited by a 2 to 3 week
treatment with imipramine (Sulser et al., 1978). In addition, we
studied whether the down regulation of ^{3}H-imipramine recognition
sites elicited by long term treatment with imipramine is associated with changes in 5HT uptake. In fact, Langer et al. (1980) had
previously suggested that ^{3}H-imipramine binding was associated with
the function of 5HT uptake mechanism.

Sulser and collaborators (1978) had reported that long term
treatment with tricyclic antidepressants causes an attenuation of
the signal amplification elicited by NE on adenylate cyclase measured in vitro on cortical slices. The data of Table 8 show that
the finding of Sulser et al. (1978) was confirmed in our laboratory and that this attenuation fails to occur if the rats had been
previously injected with 5,7-dihydroxytryptamine to destroy extensively the 5HT terminals. This destruction does not influence the
adenylate cyclase amplification elicited by NE (Table 8). From
these experiments we can draw two conclusions, first, the desensitization of brain beta-adrenergic receptors is not due to the complete inhibition of serotonergic function following the lesion of
5HT axons elicited by 5,7-dihydroxytryptamine because this lesion
per se is without consequence on the adenylate cyclase coupled to
beta adrenergic recognition sites, in contrast, the selective lesion of the brain 5HT axons impairs the capacity of imipramine to
decrease the signal amplification elicited by NE on brain adenylate
cyclase. Hence, these experiments establish that the noradrenergic
function modification elicited by imipramine does not reflect only
an action of imipramine on brain noradrenergic synapses but it is
clearly due to the activation of a transsynaptic mechanism requir-

Table 8

SELECTIVE LESION OF 5HT AXON TERMINALS PREVENTS THE IMIPRAMINE-IN-
DUCED DESENSITIZATION OF THE NE STIMULATED CYCLIC AMP ACCUMULATION
IN SLICES PREPARED FROM RAT FRONTAL CORTEX

Treatment	NE-Induced cAMP Accumulation (pmoles/mg protein/15 min)	% Increase Over Basal
Saline-sham operated	16±2.0	+82
Imipramine-sham operated	5.4±1.5*	+31
Saline-5,7-DHT	12±1.4	+74
Imipramine-5,7-DHT	13±0.8	+89

* P< 0.01 when compared to the values obtained in saline treated
sham operated rats. The basal values were 19.8±2.4; 17.0±2.5;
16.6±1.6; 14.5±1.5 pmol/mg protein/15 minutes, for saline sham
operated, imipramine sham operated, saline 5,7-DHT and imipramine
5,7-DHT, respectively. The NE concentration used in these experiments was 50 μM. The rats were treated with imipramine (10 mg/kg
i.p., twice faily for 3 weeks and were killed by decapitation 48-
72 hours after the last injection. The lesion with 5,7-DHT was obtained as described in the legend of Table 2. The cAMP accumulation in slices from rat frontal cortex was measured by radioimmuno-
assay after the incubation was carried out as described by Blumberg
et al. (1976) with minor modifications.

ing the function of serotonergic synapses. Probably, the action of
imipramine on the 5HT uptake at serotonergic synapses projects to
noradrenergic synapses, via an interneuronal system which links
serotonergic and noradrenergic transmission function (Fig. 4A).

Langer and collaborators (Langer et al., 1980 and Sette et al.,
1981) had indicated that in brain the high affinity recognition
site for ^{3}H-imipramine was located in the proximity of the neuronal
reuptake site for 5HT; in addition, their experiments had also
suggested that this site was not the recognition site for 5HT
(Ahtee et al., 1981). Hence, they had given preliminary evidence
that the reuptake for 5HT is a complex process which includes a
recognition site for 5HT, the specific carrier, and a modulatory
site which perhaps is the binding site for ^{3}H-imipramine. In order
to substantiate their proposal, we carried out experiments to study
the reuptake of 5HT in slices prepared from brain of rats receiving
for three weeks daily injections of vehicle or imipramine. As
previously mentioned in these rats the Bmax of the binding for [3]H-
imipramine is reduced (Briley et al., 1982; Brunello et al., 1982;
Kinnier et al., 1980).

The data of Table 9 show that in slices prepared from hippocampi

of rats receiving repeated injections of imipramine to down regulate imipramine binding sites, there is a more efficient 5HT reuptake than in the slices from hippocampi of rats receiving repeated
injections of saline. These results are consistent with the view
that the high affinity binding for imipramine has a regulatory role
for the uptake of 5HT (Figure 4B). Perhaps the slice preparation
still contain an endogenous effector which similarly to imipramine
reduces the uptake. When the number of ^{3}H-imipramine recognition
sites is reduced, there is a number of 5HT reuptake sites that are
deprived of an efficient regulatory site through which the endogenous effector may act. Hence, those 5HT reuptake sites which are
not anymore under a physiological inhibitory control because they
are deprived of the recognition site for imipramine necessary for
regulation contribute to the enhancement of 5HT uptake shown in
Table 9.

Table 9

REPEATED IMIPRAMINE INJECTIONS ELICIT AN INCREASE OF 5HT REUPTAKE
IN SLICES PREPARED FROM RAT HIPPOCAMPUS

Treatment	5HT Reuptake (pmol/mg prot/4 min)	%
Saline	2.30±0.22	100
Imipramine (10 mg/kg, i.p. twice daily, for 21 days)	3.80±0.29*	165

*P< 0.01 when compared to the values obtained in saline treated
rats. Each number represents the mean S.E.M.±of at least four
different determinations run in triplicate. The ^{3}H-5HT concentrations used in these experiments was 200 nM. The blank values were
determined by incubating the samples at 37°C for 4 minutes in the
presence of $5x10^{-5}$M fluoxetine, a specific 5HT reuptake inhibitor
The blank values were not significantly different in saline and
imipramine treated rats. The rats were always killed by decapitation 72 hours after the last drug injection to allow a complete-
washout of the drug, whose also partial persistency could interfere
with the 5HT reuptake measurement.

When the function of the 5HT uptake site is unrestricted, the
number of 5HT molecules which are available after each nerve impulse to activate the postsynaptic 5HT receptor will be reduced
and serotonergic transmission measured as an opening time of the
specific ion gate that functions as the transducer of the 5HT
receptor will be smaller than normal. With prolonged time periods
of 5HT uptake modification, the adaptive changes in 5HT transmission will take place and the efficiency of 5HT induced responses
will improve. The present evidence suggests that imipramine and
probably other typical tricyclic antidepressants modify brain

A.

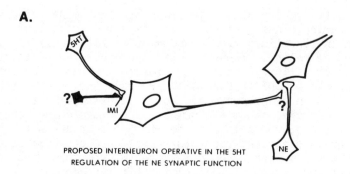

PROPOSED INTERNEURON OPERATIVE IN THE 5HT
REGULATION OF THE NE SYNAPTIC FUNCTION

B. 5HT REUPTAKE CHANNEL

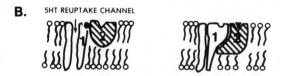

● 5HT   ▼ IMIPRAMINE AND/OR ENDOGENOUS MODULATOR
1 5HT RECOGNITION SITE
2 IMIPRAMINE AND/OR ENDOGENOUS MODULATOR RECOGNITION SITE

Figure 4

synaptic transmission by essentially two types of mechanisms. A
single injection of imipramine stimulates specific recognition
sites that inhibit 5HT uptake persistently. The short term effect
of imipramine treatment is to increase the amount of 5HT present in
synapses and therefore to increase the transducer response elicited
by 5HT in the postsynaptic membrane, as a result 5HT transmission
is facilitated. However, on protracted imipramine administration,
the number of specific recognition sites for imipramine is down-
regulated; therefore a number of 5HT uptake sites will have lost
their regulatory sites and, therefore, the capability of their
being regulated by the endogenous effector and 5HT uptake will be
increased.

At 5HT synapses, the amount of 5HT reaching the transducer will
be decreased and the responses will be reduced either in inten-
sity or in frequency. On a long term basis this reduction in the
tone of 5HT synapses causes a down regulation of beta-adrenergic
receptor number and a desensitization of their transducer system
via a collateral neuronal circuit depicted in Fig. 4A. This transsynaptic modulation of NE synapses elicited by a persistent change
in the tone of 5HT transmission, appears to be a necessity at least
for the imipramine induced down regulation of beta-adrenergic receptors because it fails to occur when 5HT axons are lesioned.
Since the absence of 5HT axons per se is not sufficient to down

regulate NE transmission, it can be argued that a fine change in
the tuning of 5HT transmission triggers the signal that attenuates
noradrenergic synaptic amplification. This attenuation is now
being proposed to have some therapeutic value in depression. If the
interpretation given above is correct, it is impossible to believe
that depression is the result of a deficiency in NE transmission.

Speculative Summary

With the realization that there are synapses with two chemical
signals (transmitter and cotransmitter) it is possible to hope that
drugs can be found that will modify synaptic function by acting on
the cotransmitter signal that tunes the gain of synapses and that
these drugs can be used to ameliorate certain symptoms of mental
diseases. Since these drugs do not act on the primary transmitter,
a number of unbearable side effects should be limited. This new
pharmacology of the cotransmitter receptor is already operative with
the benzodiazepines. These drugs have been used by billions of
patients and they have given the lowest rate of side effects ever
experienced with centrally acting drugs. The benzodiazepines act
on a cotransmitter that operates the fine tuning of GABAergic synapses. The discovery that benzodiazepines relieve anxiety because they
modulate GABA synapses has led to the study of the GABAergic mechanism as a focus for elucidating the neurobiology of anxiety. In
fact, the endogenous agonist of the benzodiazepine recognition
site modulates GABAergic synapses eliciting opposite effects of
those elicited by the benzodiazepines. The endogenous effector produces anxiety rather than relieving it. With the purification of a
neuropeptide which appears to be structurally related to the endogenous agonist that modulates GABA receptor in vivo, we have come
close to identifying a biochemical mechanism operative in
anxiety. With a better understanding of this neuropeptide (amino
acid sequence, synthesis, catabolism, receptors) we could obtain
methods to study in spinal fluid or in other body fluids whether
this anxiogenic factor is a biochemical marker of anxiety. Thus, we
will acquire the possibility to define anxiety in biochemical terms.
A similar possibility appears to be forthcoming for depression. Imipramine studies have contributed information that can be construed
to indicate that the uptake of 5HT is a regulated process. This
regulation is similar to that occurring in other membrane pumps for
ions. The acceptable pump is not only regulated by the amount of energy available but also by an endogenous effector which changes the
characteristics of the pump. Many transmitters act by modifying
pumps, gates, or channels for the passage of ions across the neuronal membranes. Likewise, uptake of 5HT appears to have a site for
an endogenous effector; this site is shared by imipramine and,
probably, other typical tricyclic antidepressants. Imipramine, like
the endogenous effector, probably inhibits the uptake of 5HT caus-
ing an increased stimulation of 5HT receptors. But, on continued
administration, the imipramine binding sites are down regulated,

reuptake of 5HT increases, and 5HT tone is reduced. Perhaps, depression is due to a functional defect of the endogenous modulator for 5HT uptake. Perhaps, there is an increase in the quantal size that is released by nerve impulses. Since the release is discontinuous there is no chance for an effective down regulation of the receptor sites. It takes two weeks to notice the relief of the symptoms of depression by imipramine because it takes that long to down regulate the number of sites of action of the endogenous effector that operates the fine tunine of 5HT uptake. We are now in the process of isolating and purifying the endogenous effect of 5HT uptake regulation which could become an important marker for depressive syndrome. It is possible to hope that depression or a group of depressions can be described in terms of a functional abnormality in the modulator of 5HT uptake.

References

Ahtee, L., Briley, M., Raisman, R., Lebrec, D. and Langer, S.Z., 1981, Reduced uptake of serotonin but unchanged ^{3}H-imipramine binding in the platelet from cirrhotic patients, Life Sciences 29, 2323-2329

Baraldi, M., Guidotti, A., Schwartz, J.P. and Costa, E.,1979,GABA receptors in clonal cell lines: a model for study of benzodiazepine action at the molecular level. Science 205: 821-823

Biggio, G., Brodie, B.B., Costa, E. and Guidotti, A., 1977, Mechanisms by which diazepam, muscimol, and other drugs change the content of cGMP in cerebellar cortex. Proc. Nat. Acad. Sci. 74: 3592-3596

Blumberg, J.B., Vetulani, J., Stawaiz, R.J. and Sulser, F., 1976, The noradrenergic cyclic AMP generating system in the limbic forebrain. pharmacological characterization and possible role in the mode of action of antipsychotics. European J. Pharmacology 37: 357-366

Braestrup, C., Schmiechen, R., Neef, G., Nielsen, M. and Petersen, E.N., 1982, Interaction of convulsive ligands with benzodiazepine receptors. Science 216: 1241-1243

Braestrup, C. and Squires, R., 1978, Benzodiazepine receptors in brain. Nature 26:1680-1683

Briley, M., Raisman, R., Arbilla, S., Casadamont, M. and Langer, S.Z, 1982, Concomitant decrease in ^{3}H-imipramine binding in cat brain and platelets after chronic treatment with imipramine. European J. Pharmacol.91: 309-314

Brownstein, M.J., Saavedra, J.M., Axelrod, J., Zamon, G.H. and
Carpenter, D.O., 1974, Coexistence of several putative neurotrans-
mitters in single identified neurons of aplysia. Proc. Nat.
Acad. Sci.71: 4662-4665

Brunello, N., Chuang, De-Maw, and Costa, E., 1982, Different synaptic locations of mianserin and imipramine binding sites. Science
215: 1112-1115

Cheng, Y.C. and Prusoff, W.H., 1973, Relationship between the inhibition constant (K_i) and the concentration of inhibitor which
causes a 50% inhibition (IC_{50}) of an enzymatic reaction. Biochem.
Pharmacol. 22: 3099-3108

Corda, M.G., Blaker, W.D., Mendelson, W., Guidotti, A. and Costa, E.,
1983, Beta-carbolines enhance shock-induced suppression of drinking
by acting on benzodiazepine recognition sites, Proc. Nat. Acad. Sci,
in press

Corda, M.D., Costa, E., and Guidotti, A., 1982, Specific proconvulsant action of an imidazobenzodiazepine (RO-15-1788) on isoni-
azid convulsions. Neuropharmacology 21: 91-94

Costa, E., DiGiulio, A., Fratta, W., Hong, J. and Yang, H.Y-T., 1979,
Interactions of enkephalinergic catecholaminergic neurons in CNS
and periphery. In: E. Usdin, I.J. Kopin, J.D. Barchas (eds):
Catecholamines: Basic and Clinical Frontiers, Pergamon Press,
Oxford, pp 1020-1025

Costa, E. and Guidotti, A., 1979, Molecular mechanisms in the re-
ceptor action of benzodiazepines. In: R. George, R. Okun and A.K.
Cho (eds.): Annual Review of Pharmacology and Toxicology 19:
Palo Alto, CA, Annual Reviews, Inc., pp 531-545

Costa, E., Guidotti, A., Hanbauer, I., Hexum, T., Saiani, L., Stine,
S., 1981, Regulation of acetylcholine receptors by endogenous
cotransmitters: studies of adrenal medulla. Fed. Proc. 40:
160-165

Costa, E., Guidotti, A., Hanbauer, I. and Saiani, L., 1983, Modulation of chromaffin cell nicotinic receptor function by opiate
recognition sites highly selective for Met^5-enkephalin-$Arg^6$$Phe^7$.
Fed. Proc., in press

Costa, E., Guidotti, A., Mao, C.C. and Suria, A., 1975, New concepts on the mechanism of action of benzodiazepines. Life
Sciences 17: 167-185

Costa, E., Guidotti, A., and Toffano, G., 1978, Molecular mechanisms mediating the action of diazepam on GABA receptors. British J. Psychiatry 133: 239-248

Cottrel, G.A., 1977, Identified amine containing neurons and their synaptic connections. Neuroscience 2: 1-18

Dale, H.H., 1935, Pharmacology and Nerve Endings. Proc.Royal Soc Med. 28: 319-332

Eccles, J.C., Fatt, P. and Londgren, S., 1956, Central pathway for direct inhibitory action of impulses in largest afferent nerve fibers to muscle. J. Neurophysiol. 19: 75-98

Guidotti, A., Gale, K., Suria, A., Toffano, G., 1979, Biochemical evidence for two classes of GABA receptors in rat brain. Brain Research 172: 566-571

Guidotti, A., Konkel, D.R., Ebstein, B., Corda, M.G., Wise, B.C., Krutzsch, H., Meek, J.L. and Costa, E., 1982, Isolation, characterization, and purification to homogeneity of a rat brain protein (GABA-modulin). Proc. Nat. Acad. Sci. 79: 6084-6088

Guidotti, A., Toffano, G. and Costa, E., 1978, An endogenous protein modulates the affinity of GABA and benzodiazepine receptors in rat brain. Nature 257:553-555

Haefely, W., Pieri, L., Polc, P., Schaffner, R., 1981, General pharmacology and neuropharmacology of benzodiazepine derivatives. Handbook of Experimental Pharmacology 55: 13-62

Haefely, W., Polc, P., Schaffner, R., Keller, H.H., Pini, L., Mohler, H.., 1979, Facilitation of GABAergic transmission by drugs. In: P. Krogsgaard-Larsen, J., Scheil-Kruger and H. Kopad (eds.): GABA Neurotransmitters, Munksgaard, Coopenhagen, pp 357-375

Hokfelt, T., Lundberg, J.M., Schultzberg, M., Johansson, O., Ljungdahl, A. and Rehfeld, J., 1980, Coexistence of peptides and putative transmitters in neurons. Advances in Biochemical Psychopharmacology 22: 1-23

Kerkut, G.A., Sedden, C.L., Walker, R.T., 1967, Uptake of Dopa and 5-hydroxytryptophan by monoamine forming neurons in the brain of Helix Aspersa. Comp. Biochem. Physiol 23: 157-162

Jeste, D.V. and Wyatt, R.J., 1982, The Understanding and Treating of Tardive Dyskinesia, The Guilford Press, New York,

Kinnier, W.J., Chuang, D.M., Costa, E., 1980, Down regulation of dihydroalprenolol and imipramine binding sites in brain of rats repeatedly treated with imipramine. European J. Pharmacology 67: 289-294

Kumakura, K., Karoum, F., Guidotti, A. and Costa, E., 1980, Modulation of nicotinic receptors by opiate receptor agonists in cultured adrenal chromaffin cells. Nature 283: 489-492

Langer, S.Z., Moret, C., Raismon, R., Dubocovich, M.L. and Briley, M., 1980, High affinity ^{3}H- imipramine binding in hypothalamus: association with uptake of serotonin but not of norepinephrine. Science 210: 1133-1136

Majewska, M.D. and Chuang, De-Maw, 1983, Benzodiazepine-induced facilitation of ^{3}H-GABA binding to GABA$_A$ recognition site: inhibition by Ca^{2+}. J. Biol. Chem., in press

Mohler, H., Bettusby, M.K., Richards, J.G., 1980, Benzodiazepine rectptor protein identified and visualized in brain tissue by photoaffinity label. Proc. Nat. Acad. Sci. 77: 1666-1670

Mohler, H. and Okada, T., 1978, The benzodiazepine receptor in normal and pathological human brain. British J. Psychology 133: 261-268

Mohler, H., Wu, J., Richards, J.G., 1981, Benzodiazepine receptors: autoradio and immunocytochemical evidence for their location in regions of GABAergic synaptic contacts. In: E. Costa, G. DiChiara and G.L. Gessa (eds): GABA and Benzodiazepine Receptors, Raven Press, New York, pp 139-146

Palkovits, M., Raisman, R., Briley, M. and Langer, S.Z., 1981, Regional distribution of ^{3}H-imipramine in rat brain. Brain Research 210: 493-498

Panula, P., Yang, H.-Y.T., and Costa, E., 1983, Coexistence of Met enkephalin-Arg^6-Phe^7 with Met^5-enkephalin and the possible role of Met^5-enkephalin-Arg^6-Phe^7 in neuronal function. In: Proceedings of a symposium entitled, "Coexistence of Neuroactive Substances" held at the NIH, in press

Paton, W.D.M., 1954, Transmission and block in automonic ganglia Pharmacological Review 6: 59-67

Raisman, R., Briley, M.A., Langer, S.Z., 1979, Specific tricyclic antidepressant binding sites in rat brain. Nature 284: 17-21

Saiani, L. and Guidotti, A., 1983, Opiate receptor mediated inhibition of catecholamine release in primary cultures of bovine adrenal chromaffin cells. J. Neurochemistry, in press

Sette, M., Raisman, R., Briley, M. and Langer, S.Z., 1981, Localization of tricyclic antidepressant binding sites on serotonin nerve terminals. J. Neurochemistry 37: 40-42

Shildkraut, J.J., 1965, The catecholamine hypothesis of affective disorders: a review of supporting evidence. Am. J. Psychiatry 122: 509-522

Shildkraut, J. J., 1978, Current status of the catecholamine hypothesis of affective disorders. In: Lipton, M.A., DiMascio and K.F. Killam (eds): Psychopharmacology: A Generation of Progress, Raven Press, New York, pp 1223-1234

Study, R.E. and Barker, J., 1981, Diazepam and (-) pentobarbital: fluctuation analysis reveals different mechanisms of the potentiation of GABA responses in cultured central neurons. Proc. Nat. Acad. Sci. 78: 7180-7184

Sulser, F., Vetulani, J., Mobley, P.L., 1978, Mode of action of antidepressant drugs. Biochem. Pharmacol.27: 257-262

Tallman, J.F., Thomas, J.W. and Gallagher, D.W., 1978, GABAergic modulation of benzodiazepine binding site sensitivity. Nature 24: 383 385

Tang, J., Yang, H.-Y.T., and Costa, E., 1982, Distribution of Met-enkephalin-Arg^{6}Phe^7 in various tissues of rats and guinea pigs. Neuropharmacology 21: 595-600

Wise, B.C., Guidotti, A. and Costa, E., 1983, Phosphorylation induces a decrease in the biological activity of the protein inhibitor (GABA-modulin) of gamma-aminobutyric acid binding sites. Proc. Nat. Acad. Sci., in press

Yang H.-Y.T., Majane, E. and Costa, E., 1981, Conversion of Met^5-enkephalin-Arg^6-Phe^7 to met^5-enkephalin by dipeptidyl carboxypeptidase. Neuropharmacology 20: 891-894

Yang, H.-Y.T., Panula, P., Tang, J. and Costa, E., 1983, Characterization and location of Met^5-enkephalin-Arg^6-Phe^7 stored in various rat brain regions. J. Neurochemistry, in press

STEROID HORMONE RECEPTORS

Paolo Marchetti,°Gigliola Sica, Clara Natoli, Maria As-
sunta Spina, and Stefano Iacobelli
Laboratory of Molecular Endocrinology and °Department
Histology, Catholic University of the Sacred Heart
Largo A.Gemelli,8 00168 Rome, Italy

INTRODUCTION

Over the past twenty years there have been significant advances
in the understanding of the biochemical mechanisms by which steroid
hormone interaction with specific receptors elicits biological
responses in target cells. A complete discussion of the available
data on this subject is beyond the scope of this paper. We will
focus our attention on the currently accepted two-step pathway
of intracellular interaction of steroid hormones.

MECHANISM OF ACTION OF STEROID HORMONES

After entering the target cell, steroid hormones interact
with cytoplasmic, high affinity, low capacity and stereospecific
receptor proteins. This association induces a steroid-dependent,
energy-requiring conformational change (transformation or activation)
of the receptor-hormone complex, which is then able to translocate
in the nucleus. The interaction of the transformed complex with
nuclear acceptor sites results in a increase in specific RNA
messengers, which elicit the cellular responses to hormones. Despite
the numerous data available on this subject, several problems are
still unsolved, e.g. receptor and/or receptor-hormone complex
activation, nuclear uptake, biosynthesis, degradation and intracellular compartmentalization of receptors. We shall review the
broader aspects of this generally accepted two-step model for
steroid action.

The steroid probably enters the target cell by passive diffusion, even if the presence of a protein-facilitated diffusion
not requiring energy at the level of the plasma membrane has

been proposed (Cake and Litwack, 1975; Zanka et al., 1981).

The cytoplasmic form of the receptor interacting with the steroid is still open to debate. Sucrose density gradient analysis has been extensively employed to demonstrate specific steroid receptors. However, many factors such as ionic strength, temperature, presence of stabilizing compounds and enzyme inhibitors are capable of modifying receptor structure, so results are difficult to compare. Recently Notides et al. (1981) demonstrated that the non-activated estrogen receptor from calf uteri has a sedimentation coefficient of 4S. In low-salt (less than 0.3M) sucrose density gradient a molecule of 8S - probably composed of four 4S units - appears (Sica et al., 1976). The cytoplasmic estrogen receptor from the human uterus shows a sedimentation coefficient of 3-4S in addition to the 8S form, probably due to the presence of proteases. The salt-dissociated form of this receptor has a sedimentation coefficient of 3.8S (Notides et al., 1972). Sedimentation values of progesterone receptor range from 3.5 to 8S (Clark and Peck, 1979). Schrader et al., (1980) have shown that progesterone receptor in the chich oviduct is a 6S dimer composed of two 4S subunits representing two distinct parts of a functional complex, even if another hypotheses on the role of these subunits has been proposed (Catelli et al., 1983). The native form of the human progesterone receptor has a sedimentation coefficient of 7S in low salt and 4S in high salt sucrose density gradient (Verma and Laumas, 1973). Cake and Litwack (1975) proposed that the 7-8S glucocorticoid receptor does not have a more important physiological role than the 4S form. The presence of an 8S and a 4S form of cytoplasmic androgen receptor has also been reported (Mainwaring and Johnson, 1980). DEAE chromatography and polyacrylamide gel electrophoresis seem to be capable of providing more precise information on the presence of various forms of the native steroid receptor, even if their biological significance has not yet been clarified (Clark and Peck, 1973; Catelli et al., 1983).

The transformation (or activation) process of cytoplasmic receptor after hormone binding has been studied by evaluating the modifications of the sedimentation coefficients, the dissociation kinetics and the affinity for polyanions and nuclei (Cake and Litwack, 1975; Notides, 1978; Milgrom, 1980; Seibert and Lippman, 1982; Catelli et al., 1983). After hormone interaction progesterone, glucocorticoids and androgen cytoplasmic receptors change from a 7.8S form to a smaller form and sediment at 4S. On the other hand, the transformed estrogen receptor (probably a dimer) shows a sedimentation coefficient of 5S (Notides et al., 1981). The transformation or activation process may be the result of an enzymatic transformation or simply a conformational change. As far as glucocorticoids are concerned, the activa-

tion process following interaction with the hormone seems to be due to
a dephosphorylation process controlled by a modulator protein present
in the plasma (Sekula et al., 1981).

The subsequent step in steroid action is the translocation of
the transformed complex into the nucleus followed by binding to DNA
and/or chromatin (Feldman et al., 1972; Puca et al., 1974). The physicochemical properties of the nuclear steroid receptors are very similar to those of activated cytosol receptors. Current data support the
view that steroid hormone receptors interact with chromatin proteins
either by catalyzing gene transcription or by blocking a repressor
protein that controls specific gene expressions (Mester et al., 1979).

After nuclear translocation, estrogen receptor content in the
nucleus drops drastically in a few hours without a parallel increase
in unfilled receptors in the cytosol. Kasid et al., (1983) recently
proposed that the processing is not a loss of nuclear receptors but
a receptor modificaiton implying a lesser solubility of the complex
which is strictly associated with DNA, and a slow dissociation rate.
This could explain why processing seemed to be a fundamental requis-
ite for estrogen action (Horwitz and McGuire, 1978). The inactivation of the estrogen receptor after interaction with the nuclear
acceptor sites seems to be due to dephosphorylation by a nuclear
phosphatase. Later the receptor is destroyed or inactivated by a
phosphorylation process in the cytoplasm (Migliaccio and Auricchio,
1981). The replenishment of cytoplasmic glucocorticoid receptors is
probably due to the reappearance of the original receptor in the
cytoplasm rather than to a new receptor synthesis (Cake and Litwack,
1975; Munck and Foley, 1976).

The single two-step model we previously examined leaves some
unexplained points, such as the possible nuclear localization of
unfilled receptors in equilibrium with cytoplasmic receptors, the
hypothesized nuclear activation of the receptor-hormone complex and
the nature of the acceptor site. Morover, we know very little about
the composition, amino acid sequence and physicochemical properties
of the various classes of steroid hormone receptors and the purification and identification of the various forms of receptors (filled
or unfilled, cytoplasmic or nuclear), the identification of the nuclear acceptor sites and the specific sequences of DNA and genes
involved in the hormonal response.

Finally, a more thorough knowledge of the mechanism of action of
steroid hormones will make it possible to choose between various
methods of analysis and receptor assay (for a review see Clark and
Peck, Seibert and Lippman, 1982).

REFERENCES

Cake, M.H., and Litwack,G., 1975, The glucocorticoid receptor, in: "Biochemical Actions of Hormones", G.Litwack, ed., Academic Press, New York.

Catelli, M.G., and Mester,J., 1983, The mechanism of action and effect of ovarian steroids, in: "The Ovary", G.B.Serra,ed., Raven Press, New York.

Clark,J.H., and Peck,E.J., 1979, Female sex steroids. Receptors and function, Springer-Verlag, Berlin.

Feldman,D., Funder,J.W., and Edelman,I.S., Subcellular mechanisms in the action of adrenal steroids, Am.J.Med., 53:545.

Horwitz,K.B., and Mc Guire,W.L., 1978, Nuclear mechanism of estrogen action: effects of estradiol and antiestrogens on estrogen receptors and nuclear receptor processing, J.Biol.Chem., 253:8185.

Kasid,A., Strobl,J.S., Greene,G.L.,and Lippman,M.E., 1983, Characteristics of a new nuclear form of estradiol receptor in MCF-7 breast cancer cells, Nature , in press.

Mainwaring,W.J.P., and Johnson,A.D., 1980, Use of the affinity label 17   bromoacetoxytestosterone in the purification of androgen receptor proteins, in : "Perspectives in Steroid Receptor Research," F.Bresciani,ed., Raven Press, New York.

Mester,J., Seeley,D., Catelli,M.G., Binart,N., Geynet,C., Sutherland, R.L., and Baulieu, E.E., 1979, Chicken oviduct nuclear oestrogen receptors: aspects of steroid hormone action, J.Steroid Biochem., 11:307.

Migliaccio,A., and Auricchio,F., 1981, Hormone binding of estradiol 17   receptor: evidence for its regulation by cytoplasmic phosphorilation and nuclear dephosphorilation. Prevention of dephosphorilation by antiestrogens, J. Steroid Biochem., 15:369.

Milgrom, E., 1980, Activation of steroid receptor complexes, in: "Biochemical Actions of Hormones", G. Litwack, ed., Academic Press, New York.

Munck,A., and Foley, R., 1976, Kinetics of glucocorticoid-receptor complexes in rat thymus cells, J. Steroid Biochem., 7:1117.

Notides,A.C., Hamilton,D.E., and Rudolph, J.H., 1972, Estrogen binding proteins of the human uterus, Biochem. Biophys.Acta, 271: 214.

Notides,A.C., Lerner,N., and Hamilton,D.E., 1981, Positive cooperativity of the estrogen receptor, Proc.Natl.Acad.Sci.U.S.A., 78:4926.

Puca,G.A., Sica,V., and Nola,E., 1974, Identification of a high affinity nuclear acceptor site for estrogen receptor of calf uterus, Proc.Natl.Acad. Sci. U.S.A., 71:979.

Schrader,W.T., Compton,J.G., Vedeckis,W.V., and O'Malley,B.W., 1980, Progesterone receptor proteins: studies of the relationship between the A and B forms, in: "Perspectives in Steroid Receptor

Research", F. Bresciani, ed., Raven Press, New York.
Seibert,K., and Lippman, M.E., 1982, Hormone receptors in breast
     cancer, in: "Clinics in Oncology", 1,3, M. Baum, ed.,
     W.B. Saunders, L.t.d., London.
Sekula,B.C., Schmidt,T.J., and Litwack,G., 1981, Redefinition of
     modulator as an inhibitor of glucocorticoid receptor activation,
     J. Steroid Biochem., 14:161.
Sica,V., Nola,E., Puca,G.A., and Bresciani,F., 1976, Estrogen binding
     proteins of calf uterus: inhibition of aggregation and disso-
     ciation of receptors by chemical perturbation with NaSCN,
     Biochem. 15:1915.
Verma,U., and Laumas,K.R.,1973, In vitro binding of progesterone to
     receptors in the human endometrium and the myometrium, Biochem.
     Biophys. Acta, 317:403.
Zanker,K.S., Prokscha,G.W., and Blumel,G., 1981, Plasma membrane-in-
     tegrated estrogen receptors in breast tissue: possible mole-
     cules for intracellular hormone level, J.Cancer Res.Cl.Oncol.
     100:135.

STEROID HORMONE RECEPTORS IN ENDOCRINE-RELATED TUMORS

Paolo Marchetti, Giovanni Scambia, °Gigliola Sica, Vittoria Natoli, and Stefano Iacobelli
Laboratory of Molecular Endocrinology and Department of Histology, Catholic University of the Sacred Heart
Largo A. Gemelli, 8 00168 Rome, Italy

INTRODUCTION

One of the main objectives in the fight against cancer is the identification of precise tumor markers which can aid in selecting the most appropriate treatment for each individual patient. This is particularly relevant with regard to tumors arising in hormone target tissues, such as breast, endometrium, ovary and prostate carcinoma or leukemia and lymphoma, which often retain the hormone sensitivity of their parent cells. Endocrine manipulations designed to modify endogenous hormone influences on some of these tumors preceeded any rational understanding of the mechanism of hormone action. The concept of steroid receptors was introduced into the scientific world for the first time about twenty years ago (Folca et al.,1961; Jensen and Jacobson, 1962). A great deal of work has been done in recent years on the mechanism of action of steroid hormones and relationships between the presence of receptors in endocrine-related tumors and both the responsiveness to therapy and the prognosis. However, it is now generally accepted that the mere presence of receptors in target cells is not sufficient to guarantee hormone responsiveness (Yamamoto and Alberts, 1976; Gehering, 1980; Lippman and Nawata, 1982).

The main purpose of this chapter is to discuss the possibility of using steroid receptor analysis to predict responsiveness to therapy and prognosis in endocrine-related tumors. Moreover, we will focus on some end-products of hormone action as markers for functional receptor sites.

BREAST CANCER

Since the pioneering observations of Cooper (1835) on the relationships between the phases of the menstrual cycle and breast
cancer growth, Schinzinger (1889) on the importance of oophorectomy
in premenopausal women with breast cancer and Beatson (1896) on
mammary tumor regression in premenopausal patients after bilateral
oophorectomy, the evidence accumulated has proven that this type
of cancer, like other tumors, requires hormonal stimuli for onset
and progression. It has also been ascertained that the regression
of these cancers may be achieved by endocrine manipulations. Both
ablative (i.e. oophorectomy, adrenalectomy and hypophysectomy) and
additive (i.e. estrogens, androgens, progestins, corticosteroids,
and, more recently, antiestrogens) endocrine therapy induce tumor
regression in some patients with advanced breast cancer. However,
only 25-30% of unselected patients respond to endocrine therapy
(A.M.A., 1960; Stoll, 1972). With the aim of increasing the percentage of responders, several clinical criteria, including menopausal status, disease-free interval, site of visceral metastasis,
lymph node involvement, and response to previous endocrine therapy
have been proposed to predict the hormonal responsiveness of the
tumor (Stoll, 1969). However, a more rational approach to therapeu-
tical choice is constituted by the measurement of steroid hormone
receptors in the tumor.

Two methods are used routinely for steroid receptor determination. The dextran-coated charcoal (DCC) assay gives a measure
of both the binding capacity and the dissociation constant. The
sucrose gradient method separates the various forms of steroid receptors, characterizing certain molecular properties of receptors.

Estrogen and progesterone receptors and response to endocrine
therapy

Studies correlating endocrine-induced remission in breast
cancer and estrogen receptor (ER) content of the tumor have been
recently reviewed by De Sombre (1982) and Seibert and Lippman (1982).

Despite the presence of differences in ER assay methods, in
patient populations and in endocrine treatments used on approximately 2,000 women studied by different authors, 55% of patients
with tumors containing ER (ER+) achieved objective remission from
endocrine therapy, while patients whose tumors were classified as
ER negative (ER-) showed a lower response rate (7%) to this kind
of therapy.

Although the use of ER analyses has substantially increased
the percentage of patients responding to endocrine therapy, it has
been established that the merely qualitative concept of ER+ and/

or ER- tumor is inadequate in predicting the response to endocrine
treatment. The potential significance of the quantitative ER content
in breast cancer has also been proposed (Jensen et al., 1975 a;Jensen
et al., 1975 b). Based on studies utilizing DCC assay for ER determination (Leclerq et al.,1975; McGuire et al., 1975 b; Lippman and
Allegra, 1978; Dao and Nemoto, 1980; Lippman and Allegra, 1980;
Osborne et al.,1980; Paridaens et al., 1980),further improvement in
prediction of responsiveness to endocrine therapy has been obtained.
If the ER concentration is high (more than 100 femtomoles (fm)/mg
protein), the response rate is 67-80%, for lesser quantities of ER,
around 40% and if receptor levels are very low (less than 3-10
fm/mg protein), the response rate is only about 9-12%.

Wittliff and Savlov (1975) suggested that the sedimentation
characteristics of ER may be predictive of the patient's response
to endocrine therapy. Using sucrose gradient separation of specific
estrogen binding components, Wittliff (1980) and Wittliff et al.
(1982) reported that sedimentation coefficients of ER in cytosol
fractions of primary and advanced breast cancer permit a good prediction of clinical responsiveness. 33 of 34 patients with tumor
containing specific steroid-binding components migrating at either 8S
or both 8S and 4S responded to endocrine therapy, while 4 of 23 patients with tumors containing only the 4S component showed an objective remission after endocrine therapy. In addition none of 44 patients with tumors in which not detectable ER were measured responded.
Although analogous evidence is reported (Mc Carty et al., 1980),
there is disagreement as to the clinical value of different receptor
forms (Dao and Nemoto, 1980; De Sombre, 1982; Wittliff et al.,
1982).

A possible way to ameliorate the ER positive patient's selection
could be represented by the knowledge of the relationships between
the receptor status and other clinical parameters, including body
weight, menopausal status, site of metastasis and certain pathological characteristics of the tumor (for review, see Seibert and
Lippman, 1982).

However, the receptor-steroid interaction is only the first
step in a multisequential mechanism and it is possible that the lack
of therapeutic response of some ER+ tumors could be ascribed to defects in steps beyond the initial binding of the steroid to the receptor. Major stress is currently being focused on the identification of end-products of hormone action which can more accurately
identify estrogen-dependent breast cancer growth.

The synthesis of progesterone receptors (PR) is unequivocally
regulated by estrogens in the endometrium (Milgrom et al., 1973),
in normal mammary epithelium (Pollow et al., 1977) and in human
breast cancer cells in culture (Horwitz et al, 1975; Horwitz and
McGuire, 1978). Since PR could represent an indicator of functioning

ER, the simultaneous measurement of ER and PR has been proposed
(Horwitz et al., 1975). The results of studies performed in several
laboratories (Matsumoto et al., 1978; Brooks et al., 1980; Lippman
et al.,1980; Manni et al.,1980; King, 1980; Skinner et al., 1980;
Young et al., 1980) show that patients with tumors containing both ER
and PR have a higher response rate (75%) to endocrine therapy than
patients of the other groups. Patients with ER- PR- tumors show the
lower response rate (10%), as it could be expected. Few patients
apparently have PR but not ER and nearly the half respond to endocrine treatment. However this group is very small (approximately
4% of the total cases) and it can reflect inaccuracies in the assay
procedure. Finally, 34% of patients with ER+ PR- tumor show a positive response. There are some explanations of the response observed
in these patients.

Even though there is a strict correlation between ER concentration and presence of PR (Horwitz and McGuire, 1975; McGuire and
Horwitz, 1978), Osborne et al., have reported clinical data confirm-
ing the usefulness of PR assay in addition to a quantitative ER
determination, instead of the ER determination alone.

Several hypotheses to explain the lack of an exact correlation
between response to hormone treatment and steroid hormone receptor
concentration have been formulated (for review,see Seibert and
Lippman, 1982).

It is to be remembered that receptor-hormone interaction is
followed by a long series of events which can be altered. These alterations could represent a phenotypical trait peculiar to the neoplastic transformation of the cell. Defects in the process of nuclear translocation, transcription or translation are described in
some variant cells of human breast cancer (Nawata et al., 1971 a,b;
Lippman and Nawata, 1982).

The evaluation of the clinical response must also be taken into
account. The stabilization of the disease (i.e.no change) is generally considered a lack of response, even if it can be beneficial
to the patient, because the arrest of a progressive disease is
often accompanied by increased well-being and relief of pain.

It is difficult to evaluate these subjective remissions and
to know if they are related to the efficacy of the endocrine therapy.

Cellular heterogeneity of the tumor may also influence receptor
evaluation (Allegra et al., 1980; Fidler and Hardt, 1981). The breast
cancer cell population may be composed of ER+ and ER- elements, as
observed by Nenci et al. (1976), using the immunofluorescence technique. It is actually not clarified if the ER- and hormone-independent cells derive from a mutation of ER+ cells or if the two dif-

ferent cell clones are simultaneously present at the onset of the disease. In this latter case, the hormone-sensitivity of the tumor should be due to the presence of a higher number of ER+ cells. The selective hormone-dependent cell death can result in the appearance of an ER- tumor. Morover, the sample for receptor assay can derive from an area which is not representative of the overall cell population. For this reason a higher receptor concentration in a few steroid receptor-positive cells could result in a positive assay even in the presence of a large number of steroid receptor-negative cells. Obviously the classification of this tumor as receptor-positive lacks any clinical value.

Great importance has been given to the examination of the specimen for the receptor assay. Poulson (1981), in an analysis of 199 samples of breast cancer obtained by surgery, observed that approximately 10% were composed of non neoplastic tissue. The presence of necrotic areas may lead to an underestimation of the receptor levels (expressed as fm/mg total tissue protein); in fact necrotic tissue contains denatured and no dosable receptors, while total protein content is due to the proteins present in the whole tissue assayed. It is interesting to note that the noticeable reduction of ER in the primary tumors of patients who were treated with radiotherapy before surgery in comparison with untreated patients (Rochefort et al., 1980) could be due to a decrease of the cell number in the tumor after radiotherapy and independent of a real reduction of receptor levels (Noel et al., 1981). In fact, even if the ER content of a single neoplastic cell is not modified, the total ER content of the tumor could be lower as the results are expressed in fm/mg total tissue protein (i.e. tumoral+peritumoral).

The variation in ER content of different metastatic sites can also be relevant (Leake et al., 1981). This may explain the regression of a specific lesion after endocrine treatment, in a patient classified ER- on the basis of ER determination made on another lesion.

The lack of response to endocrine therapy in a possible hormone-dependent ER+ tumor can occur if hormone treatment is inadequate to reduce the circulating hormones or if surgery only partially removes the endocrine tissue. For example, 10-15% of patients who do not respond to castration respond to subsequent adrenalectomy (Lippman, 1977). On the other hand, administered exogenous hormones can be inactivated by enzymes present in the tumor tissue (Adams and Wong, 1969).

Breast cancer cells contain receptors other than ER and PR (Maass et al., 1975; Lippman et al., 1977). Androgens, glucocorticoids, prolactin, insulin and thyroid hormones are known to affect breast cancer cell proliferation via specific receptors.Although

the effects of the complex interactions among different hormones in
the tumor tissue are unknown, it is possible that the unresponsiveness
of tumor containing both ER and PR depends on its modified sensitivity
to other hormones. The responsiveness to different forms of endocrine
therapy can be not directly correlated with the presence of ER, because different endocrine treatments may act through mechanisms other
than ER. On the other hand a specific endocrine therapy modifying
the hormonal environment of ER- tumor may be effective.

Finally, the presence of methodological artifacts must be considered (Lippman and Thompson, 1979). Steroid receptor is a thermo-
labile protein and errors in transport, handling and shelving may
destroy it. The assay temperature or the presence in the incubation
medium of stabilizing compounds such as molybdate or glycerol is
important. The foaming and the subsequent oxidation of receptor
during homogenization procedures should be carefully avoided (Smith,
1980). Another crucial factor is represented by protein concentration.
Poulsen (1981) has recently found that below 1-3 fm cytosolic pro-
tein/ml on doubling the cytosol protein concentrations the specific
activity of receptor increases proportionally. Then low protein concentration may produce a false negative tumor. Concentration and
time of exposure to dextran-coated charcoal also constitute a relevant point, because they may reduce the protein concentration (Poulsen, 1981) or increase the dissociation of the labelled hormone from
receptor binding sites (Hahnel and Vivian, 1975). The possible presence of contaminating plasma sex hormone binding globulin and albumin
should be considered in the choice of ligand so as to avoid an underestimation of receptor content (Chamness and McGuire, 1979). One
source of error may be the concentration of unlabelled competitor
used in DCC assay. Non specific low affinity high capacity binding
is measured in terms of the binding of the labelled ligand in the
presence of an excess of unlabelled ligand at a concentration capable
of saturating the specific component. If very high concentrations
of competitor are used, the non labelled ligand competes for binding
even to components of relatively low affinity, resulting in an increased number of specific binding sites. To avoid this error it is
necessary to use a concentration of competing ligand not greater
than   50-100 times the maximal concentration of labelled ligand
employed (Chamness and McGuire, 1975).

Receptors could be preferentially localized in the nucleus
(Panko and McLeod, 1978), leading to false negative results on the
basis of a cytosolic assay. It is also important to note that the
most common DCC assay employed for steroid receptor analysis is
able to measure only the unoccupied sites. This may become particularly relevant in the case of high plasma levels of endogenous
or administered hormones (Feherty et al., 1970). This observation
can explain, almost in part, the differences in ER concentrations
observed between pre and postmenopausal women. In fact, premeno-

pausal women have tumors with lower ER levels and show a higher incidence of ER- tumors than postmenopausal patients (Wittliff et al.,
1971; Braunsberg et al., 1974; McGuire et al., 1975 a; Lippman and
Allegra, 1978; Saez et al.,1978). However the presence of a great
number of occupied receptor sites in premenopausal women is not
sufficient to explain these findings. Utilizing an exchange assay,
which measures the total receptor content (i.e. occupied and unoccupied sites), Sakai and Saez (1976) and Namura et al. (1977) demonstrated the presence of estrogen bound to the receptor both in pre-
and in postmenopause, in this case estrogens being of extraovarian
origin . It is of interest to note that in different mammalian species estrogens increase the levels of both ER and PR in target organs, while progesterone induces opposite effects (Clark and Peck,
1979). In postmenopausal women the lack of progesterone induces a
not counteracted ER stimulation by estrogens so that high ER levels
are observed. In premenopausal women the high level of circulating
progesterone in the luteal phase of the menstrual cycle could explain the ER and PR reduction and a relative decrease of PR in
the cytoplasm due to its nuclear translocation. Receptor assay in
this case may give results which do not express the real hormone
sensitivity of the tumor (ER- PR- or ER+ PR-). On the basis of these
observations Saez et al. underlined the importance of evaluating
the receptor assay in premenopause in relation to plasmatic progesterone level. Conversely, in postmenopausal patients low levels
of circulating estrogens may be sometimes unable to stimulate PR
synthesis. Recently, Bloom et al. (1980) described the appearance
of PR after two days treatment with diethylstilbestrol in a breast
cancer metastasis of a postmenopausal patient previously classified
as ER+ PR-. In addition, similar effects are produced by the antiestrogen tamoxifen, which stimulates PR synthesis in breast cancer
cells (Horwitz et al., 1978) and breast tumor in vivo (Namer et al.,
1980). In fact, priming of PR by tamoxifen can be utilized to identify amon ER+ patients those who are more likely to respond to endocrine therapy. However, extensive clinical trials are necessary to
confirm the clinical value of these observations.

Despite these numerous possibilities of false-positive or false-
negative receptor tumors, an accurate assay and a correct interpretation of clinical and laboratory data favour reliable results.

PR does not seem to be a perfect indicator of tumor endocrine
responsiveness. Approximately 25% of patients with ER+ PR+ tumors
failed to show objective improvement following endocrine therapy,
whereas definite response is present in patients with ER+ PR- tumors.
This discrepancy could be explained by the fact that in neoplastic
tissues the presence of one estrogen-induced effect, such as the
stimulation of PR synthesis, does not imply the persistance of all
other effects, such as the regulation of tumor growth. The absence
of an exact correlation between hormone receptors and response to

endocrine therapy suggests the necessity of detecting new markers
to evaluate the hormone sensitivity of tumors.

New markers of estrogen–dependent tumor cell growth

Several assays have been proposed to reach a better selection
of breast cancer patients, such as nucleoside incorporation (Nicole
and Saes, 1978), tumor peroxidase activity (Anderson et al., 1975),
tumor stem cell assay (Hamburger et al.,1978), and 17 β -hydroxyster-
oid dehydrogenase activity (Lubbert and Pollow, 1978).

Recently, Westley and Rochefort (1979) demonstrated the presence
and inducibility of a secreted protein with a molecular weight of
46,000 (46K) in a cell line derived from human mammary carcinoma
(MCF-7). The induction of this protein was specific for steroids
capable of interacting with the estrogen receptor; furthermore,
the tamoxifen, capable of blocking the estrogen–induced cell proliferation, was equally able to inhibit estrogen induction of the
synthesis of the 46K (Westley and Rochefort, 1980).

In our laboratory, we have made similar observations on a
protein of M.W. 51,000 (51K) in human breast tumor (Natoli et al.,
1981) and in a highly estrogen–dependent CG-5 cell line (Iacobelli
et al.,1983; Natoli et al.,1983). The results obtained show that
near maximal stimulation of the rate of 51K synthesis is obtained
with 1nM 17 β estradiol. Diethylstilbestrol has a similar effect,
whereas estrone is approximately ten times less potent. Other steroids tested were completely ineffective in the induction of the
51K, suggesting that only those steroids able to interact with
ER are capable of altering the synthesis of 51K. Moreover, while
tamoxifen completely inhibits the estradiol–induced increase of
51K at a molar ratio of 1:100, it is totally unable to stimulate
51K synthesis at all concentrations used. The 51K is present and
has the same characteristics of inducibility in freshly obtained
human breast cancer tissue (Iacobelli et al.,1981).

This protein shows some interesting properties in its application to tumor hormone responsiveness prediction. Unlike PR, whose
synthesis is enhanced by estrogens and antiestrogens, estrogen-
stimulated increase of 51K is completely reversed by antiestrogens
in a way which parallels the effect on cell proliferation. Therefore,
the 51K may be a useful indicator of estrogen–dependent breast
cancer cell proliferation. In addition, the synthesis of 51K is not
confined to or acquired only by cells in long-term culture, but even
takes place in primary breast cancer tissue explants. Finally,
because the 51K is released (possibly secreted) into the culture
medium of malignant cells only (Iacobelli et al.,1981), it may
also be used as an early indicator of the onset as well as the
spread of breast tumor. The recently developed hybridoma technique

to produce an anti-51K monoclonal antibody is currently being exper-
imented with in our laboratory.

ER, prognosis and response to chemotherapy

The concentration of ER in neoplastic cells is an important biochemical marker which in itself is capable of predicting the natural
course of the tumor development, independently of any other prognostic factor, such as size and location of the primary tumor, presence
of lymph nodes or visceral metastasis and degree of histological
differentiation (Knight et al., 1977).

ER- patients have a more frequent recurrence with a shorter
disease-free interval than ER+ patients. This observation is perfectly consistent with present knowledge of the natural history of
breast cancer. The normal breast gland contains ER, even if the very
large contaminating presence of fat cells sometimes makes assays
difficult to interpret (Leclercq and Heuson, 1977). So, when a
neoplastic transformation occurs in these cells, it is conceivable
that these may lose many of their biosynthetic functions, including
the synthesis of ER. The greater the neoplastic transformation,
and therefore more advanced the loss of the differentiated status,
the easier is the loss of cell ability to synthetize ER. In this
case the absence of ER could mark a faster and more aggressively
growing tumor. In fact several authors have reported that tumors
more highly differentiated tend to be very often ER+ with respect
to those less differentiated which tend to be ER- (Heuson et al.,
1975; Rosen et al, 1975; Nicholson et al., 1981). The DNA synthesis
is enhanced in ER- tumor cells with respect to ER+ cells (Meyer et
al., 1977; Silvestrini et al., 1979); moreover, using flow cytometry,
it has been demonstrated that ER- tumors contain more cells in S
phase of the cell cycle than ER+ tumors (Kute et al., 1981).

These findings have immediate clinical implications. It is well
known that chemotherapy is more effective in tumors with high replication rate. The response rate to chemotherapy is higher in less
differentiated and more aggressive ER- tumors (Lippman et al., 1978;
Jonat et al., 1980). However, Bonadonna et al. (1981) have not confirmed these results and found that ER status does not predict the
response to a variety of cytotoxic regimens.

Receptor analysis is of value also in primary breast cancer
to predict the clinical course of the disease and to choose the
best adjuvant therapy. Patients with ER- primary breast tumor
have earlier recurrences with respect to ER+ patients (Knight III
et al., 1977; Griffiths et al., 1978). Similar observations have
been made in relation to the survival (Maynard et al., 1978).
Since the patients with ER- tumors have a poorer overall prognosis,
Knight III et al., 1981 proposed to utilize the receptor assay

in primary breast cancer to identify at the beginning a subset of high risk patients. A pilot adjuvant chemotherapy trial for stage II breast cancer has indicated that it is useful to treat very aggressively patients with ER- tumors even in the absence of other clinical unfavorable clinical parameters (Knight III et al., 1981). On the contrary Bonadonna et al.,1981 did not find differences in relapse free interval in the ER+ and ER- patients both in pre and in postmenopause, using an adjuvant polychemiotherapy.

Moreover, the primary tumor receptor content is a reliable predictor of the responsiveness to a subsequent hormone treatment in a disease recurrence (Block et al., 1978; De Sombre and Jensen, 1980; Jensen et al.,1980; Nomura et al.,1980; De Sombre, 1982). In untreated postmenopausal ER+ patients there is a good correlation among subsequent assays, independently of the time interval. ER- tumors remain negative or present low receptor levels (Paridaens et al.,1980).

However, a progressive change in hormone responsiveness and receptor status can be observed in some neoplasias. The percentage of ER+ PR+ tumors at the time of the first recurrence is similar to that found in the primary tumor, but in preterminal stages of the disease only about 20% of cancers are ER+ (Nomura et al., 1980). Whether this sharp decrease in hormone receptor content or in hormone responsivity is due to a progression of the malignancy (i.e., the prevalence of more undifferentiated more aggressive ER- cells) or to a selective killing by therapy of the ER+ cells is not currently understood.

As very often in advanced disease the amount of tissue is not enough for receptor assay, data from assay performed on the primary tumor are necessary. However, when possible, re-biopsy of accessible tumor tissue should be performed during the course of the disease to monitorize the receptor status of the tumor.

ER+ responders to endocrine therapy have a longer survival than ER+ non responders and ER- patients. Furthermore, no significant differences are demonstrated between the two latter groups (Nomura et al.,1980). Singhakowinta et al.(1980) reported that the survival time calculated from the onset of recurrence is longer for ER+ patients, but this difference is not significant if the survival time is calculated from the onset of chemotherapy. Blamey et al. (1980) showed no significant differences in survival in a group of 200 patients undergoing simple mastectomy. The survival data now available do not make it possible to draw definite conclusions (De Sombre, 1982; Seibert and Lippman, 1982).

ENDOMETRIAL CANCER

Surgery and radiation therapy are the primary treatment modalities for patients with adnocarcinoma of the endometrium. However, a significant number of patients have disseminated disease and require systemic therapy. Although different cytotoxid chemotherapies are available (Carbone and Carter, 1974; Donovan, 1974; Bruckner and Deppe, 1977; Cohen et al., 1977; Deppe et al., 1980), the advantage that endocrine treatment offers is quite evident; an elevated level of cancer-killing activity is reached with hardly any toxicity to normal tissue.

Estrogens and progesterone display an antagonistic action in the endometrium (Lerner, 1964). For this reason many investigators have postulated that prolonged exposure to estrogen stimulation unopposed by progesterone can lead to a series of lesions including hyperplasia and carcinoma (Learner, 1964; Takamizawa and Sekiya,1978). As far as actual knowledge is concerned , it can be established that in non physiological conditions characterized by prolonged estrogen exposure not counterbalanced by progesterone, the estrogens can act as tumor promoters (Iacobelli et al.,1983b). These observations constitute the theoretical basis of progestin and antiextrogen use in endometrial adenocarcinoma treatment.

While progestin therapy was introduced nearly twenty years ago (Kelley and Baker, 1961; Sherma, 1966), the mechanisms by which these compounds produce regression have not yet been fully clarified. There is some evidence that progestins act directly on the cancer cells through interaction with specific receptors, rather than indirectly by regulation of gonadotropins (Anderson, 1972; Bonte, 1972; Nordqvist,1973; Stoll,1973).Objective response rate ranging from 30% to 50% after progestin treatment of advanced endometrial adenocarcinoma has been reported by several authors (Kelley and Baker, 1961; Bonte et al.,1968; Richardson and Mac Laughlin,1978; Young, 1979; Bonte,1980; Piver et al.,1980). Even if conclusive informations on progestin adjuvant therapy are not available yet (Lewis et al., 1974; Swenerton,1982), progestins are largely administered after primary treatment of endometrial adenocarcinoma with surgery and radiotherapy.

To counteract the effects of estrogens on cell growth and division, in 1974 the use of tamoxifen in the endometrial cancer treatment has been proposed (Carbone and Carter, 1974; Donovan, 1974). The available results (Swenerton, 1980; Broens et al.,1980; Bonte et al.,1981) attribute to tamoxifen a percentage of objective responses similar to that of progestins (about 30%).

From the evidence that the selection of patients to undergo hormone therapy is still largely empirical, the importance of estab-

lishing precise criteria which can indicate tumor responsiveness
to endocrine treatment clearly emerges. In fact the immediate ini-
tiation of a suitable treatment and, in cases of definitely deficient responsiveness, the avoidance of useless treatment for many
months is of paramount importance. Among the clinical features which
seem to exert a relevant predictive role, there are the disease-free
interval, the tumor grade and the age of patients (Kohorn, 1976;
Smith, 1978). However these parameters are not sufficient for an accurate patient selection (Malkasian et al., 1971; Rozier and Underwood,
1974; Iacobelli et al., 1982). In analogy to breast cancer, the presence and the function of steroid receptor in endometrial cancer may
represent a valuable tool in the choice of   treatment.

Estrogen and progesterone receptors and response to endocrine therapy

     Methodological differences and assay variations may make barely
comparable the ER and PR concentrations reported in the literature
for endometrial adenocarcinoma. A critical survey of the problem
has been recently published by U.I.C.C. (Richardson and MacLaughin,
1978).

     Crocker et al. (1974), Pollow et al. (1975) and Martin et al.
(1979) found ER in all endometrial adenocarcinomas studied. Similarly
Robel et al. (1981) demonstrated the presence of cytoplasmic and
nuclear ER in 36/37 cases examined, while other authors have found
ER in a variable percentage of cases, ranging from 45% to 85% (Terenius et al., 1977; Janne et al., 1979;Ehrlich and Young, 1980). No
qualitative differences were found between ER in normal and neoplastic endometrium (Pollow et al., 1980). The ER levels reported in the
various studies are very variable ranging from 0 to about 2,000 fm/
mg protein, but generally they were similar to those found in normal
proliferative endometrium (Pollow et al., 1980; Robel et al., 1981).
The higher ER levels were found in well differentiated carcinomas
(Terenius , 1977; Janne et al., 1979), although Pollow et al. (1977
b) reported an opposite trend.

     Our results (Iacobelli et al., 1980 a) indicate that PR are
demonstrable in 61% of adenocarcinomas studied with an average
concentration of 188 fm/mg cytosol proteins (range: 25-899 fm/mg
protein). Similar values for PR in endometrial carcinoma have been
reported (Janne et al., 1979; Martin et al., 1979), even if other
investigators found higher levels (Ehrlich and Young, 1980; Pollow
et al., 1980). As can be expected PR concentrations are more elevated
in well differentiated than in poorly differentiated carcinomas.
For example, Young et al. (1976) found PR in 58% of patients
studied and studying the relationship between histological grade
and PR levels they showed that 85% of the grade I, 54% of the grade
II and 22% of the grade III tumors were PR+ (i.e., PR concentrations
were higher tha 50 fm /mg protein). The presence of PR seems to be

a useful tool in the choice of therapy. In fact, Ehrlich and Young, 1980, reported objective response following progestin treatment in 6/6 PR+ patients and only in 2/6 PR- patients. Martin et al. (1979) also observed a similar correlation.

The simultaneous presence of both ER and PR has been found in a percentage ranging from 41 to 80% (Pollow et al., 1977 b; Janne et al., 1979; Martin et al., 1979; Ehrlich and Young, 1980). The steroid receptor determination at the time of first diagnosis seems well correlated with ER and PR status of the metastasis (Gurpide, 1981), suggesting the usefulness of the receptor level measurement on the tisuue obtained by surgery.

Progestin treatment induces a profound modification in the ER and PR content of the tumor. Since the progestins have an inhibitory effect on both ER and PR (Milgrom et al., 1973; Hsueh et al., 1975; Janne et al., 1979; Ismaa et al., 1979), it would appear necessary to investigate further the long-term effects of this kind of treatment. The prolonged administration of these compounds may reduce the hormone-dependent cell responsiveness and, of course, facilitate the overcoming of hormone-independent cells. Given that antiestrogens and progestins act through different mechanisms including receptors, and that antiestrogens like estrogens induce PR synthesis in endometrial cancer cells, it seems reasonable to propose the use of tamoxifen and medroxyprogesterone acetate, administered sequentially, for the therapy of advanced endometrial cancer. The experimental data available call for the evaluation of the therapeutic efficacy of the combined treatment in future clinical trials. These studies should be built up in such a way that tumor cells are first "primed" with antiestrogen and then killed by progestin.

Markers of functional receptor sites

The "in vitro" response to progestins may represent a useful indicator of the PR+ tumor's responsiveness to hormone therapy. Organ cultures have been successfully used by several investigators to study the effects of hormones on neoplastic tissue growth (Nordqvist, 1964; 1970). We have studied the effects of various steroid hotmones and antiestrogens on short-term organ culture of human endometrial carcinoma (Iacobelli et al., 1980 a). Among the 11 PR+ carcinomas, 7 proved to be sensitive to medroxyprogesterone acetate "in vitro", while for the other 4,despite the presence of appreciable quantities of receptor, there was no corresponding decrease of macromolecular synthesis. The 7 carcinomas without receptors proved to be insensitive to the inhibitory effect of medroxyprogesterone acetate on the incorporation of labelled precursors. Assuming a reduction of macromolecular synthesis of at least 5% as significant, the percentage of explants classifiable as positive is about 55%. This value is

very close to the percentage of carcinomas which effectively respond
to progestin treatment. This kind of test is not conceptually
linked to the hypothesis that hormone action on cancer cells is
necessarily mediated by receptor hormone interaction. In fact, the
presence of PR could simply represent a biochemical indicator of the
cell sensitivity to progestins and only as such be correlated with
hormone therapy. In other words, in cases of pharmacological doses
of hormone, it cannot be said that since the cell produces the
receptors it therefore responds, but it could be that it has, as its
characteristics, both the "responsiveness" and the "production of
receptors" (the only link could be the degree of cell differentiation). It should be added that using the measurement of PR alone,
it is possible to have falsely positive results and this could be
due either to a defect in some mechanism following receptor-hormone
interaction or the above mentioned independence of the two phenotypical characteristics (i.e. the cell has the "receptor synthesis"
gene and does not have the "sensitivity" gene).

It has been proven that, during the cycle, with the increased
progesterone levels in the serum and in the endometrium, there is
a significant decrease in the quantity of ER and an approximately
10-fold increase in 17- - hydroxysteroid dehydrogenase (17 β HSD)
activity (Tseng and Gurpide, 1975a; 1975b). Gurpide (1981) showed
the induction of this enzyme in some endometrial cancersafter 2-10
days treatment with medroxyprogesterone acetate. However, the data
which would allow a correlation between the short-term responsiveness to progestins and the effectiveness of the therapy are still
missing.

On the basis of the evidence that estrogens induce the rapid
synthesis of specific proteins in the animal uterus, we have studied
the human endometrium (Iacobelli et al.,1981a), with the dual
purpose of first seeing whether estrogen-induced proteins are
more general event of estrogen action independent of the species
to which the responsive cell belongs and, secondly, exploring the
use of these proteins as end products of estrogen action, i.e;
as markers of estrogen-stimulated cell growth and proliferation
in potential hormone-responsive tumors, in conjunction with other
already existing indicators. The human endometrial tissue fragments
taken during the secretory phase of the menstrual cycle respond
to estrogen stimulation by increased synthesis of a specific protein
with molecular weight of about 55,000 daltons. Although the exact
biological function of the estrogen-induced protein is not yet
known, our recent studies at the Weizmann Institute (Kaye et al.
1981) have demonstrated that the major constituent of rat uterine
estrogen-induced protein is the BB-isozyme of creatine kinase (CK-BB).
When CK-BB activity of human endometrium is assayed throughout the
menstrual cycle, an increase is seen during the late secretory

stage. The CK-BB could play a role in energy mobilization for estrogen-stimulated growth and cell division. In addition, the finding of an easily measurable enzymic activity for the estrogen-induced protein makes it extremely more convenient as marker protein for exploring the estrogen dependence of normal and tumor cells.

In conclusion, we believe that, even though a definite solution of the problem of identifying the hormone sensitivity of a tumor has not yet been found, some of the tests discussed already represent not only an indispensable diagnostic tool but also a successful starting point for further progress.

HUMAN LEUKEMIA AND LYMPHOMA

It has been recognized that glucocorticoids have profound metabolic inhibitory effects on the cells of lymphoid tissue, which result in cytolysis in sensitive species (Moon, 1936; Selye, 1937; Dougherty, 1952; Dougherty et al., 1964; Claman, 1972). Based on these properties, glucocorticoids have been advantageously employed first alone (Vietti et al., 1965; Livingstone, 1970; Claman, 1972) and now in combination polychemotherapy together with cytotoxic drugs (Goldin et al., 1971; Spiers et al., 1977; Bird, 1979; Mandelli et al., 1980; Shaikh et al., 1980; Vaughan et al., 1980) in the treatment of cancer arising in lymphoid tissue. However, their use may be associated with serious side effects (Kjellestrand, 1975). Consequently, it would be of value to discriminate in advance glucocorticoid responsive and non responsive cases, and restrict the use of glucocorticoids to beneficial situations.

In 1973 (a) Lippman et al. demonstrated that the leukemic lymphoblasts of 22 patients with acure lymphoblastic leukemia (ALL) contained detectable levels of glucocorticoid receptors (GR). In a subsequent study, Lippman et al. (1975) reported that the cells of patients with acute myelogenous leukemia (AML) had appreciable GR levels in only 3 out of 16 cases. In independent studies, Gailani et al. (1973) found that the cells of 4 out of 4 lymphosarcoma patients, 3 out of 3 ALL patients, 2 out of 6 AML patients, 0 out of 8 patients with chronic lymphocytic leukemia (CLL) and 0 out of 2 patients with chronic myelogenous leukemia (CML) contained appreciable GR levels. Furthermore Terenius et al. (1976) reported that the lymphocytes of only 17 out of 27 patients with CLL contained significant levels of GR. These earlier studies on the occurrence of GR in various types of leukemia were performed usin a cytosol assay. More recently, using a whole cell assay, several laboratories have quantified GR in a variety of human leukemias including ALL (Crabtree et al., 1978; Homo et al., 1980; Kontula et al. 1980; Mastrangelo et al.1980; Bloomfield et al. 1981; Daniel et al.1981b; Marchetti et al. 1981), AML (Crabtree et al., 1978; Iacobelli et

al.,1978; Sloman and Bell, 1979; Kontula et al., 1980; Crabtree et
al, 1981; Iacobelli et al., 1981 b), CLL (Homo et al., 1975;Crabtree
et al., 1978; Stevens et al., 1978; Homo et al., 1978; Kontula et al.,
1980; Iacobelli et al., 1981 b) and the Sezary sindrome (Schmidt and
Thompson, 1979).

So far, however, these studies have produced conflicting results.
GR have been determined in cytosol, nuclei and in whole cells. Therefore some of these discrepancies could be related to the different
methods used to measure the receptors. On the basis of several reports
(Crabtree et al., 1978; Duval and Homo, 1978; Homo et al., 1978) the
cytosol assay seems unreliable for the measurement of GR in leukemic
cells. Indeed, normal as well neoplastic lymphocytes contain a very
thin rim of cytoplasm and they may have a different susceptibility
to breakage during homogenization procedures, moreover, the unfilled
receptor can be preferentially localized in the nucleus of the cell
(Neifeld et al., 1977). Thus, an assay which measures only the cytoplasmic receptor concentration may lead to an underestimation of cellular GR content. On the other hand, by using the whole cell assay,
some of these problems can be overcome and appropriate measurement
of total (cytoplasmic + nuclear) receptors can be obtained. A major
technical problem associated with this assay technique is the occurrence of a high non specific binding, which, however, may be removed
by accurate post-incubation cell washing (Kontula et al., 1980).
If the total receptor sites/ cell are calculated as the sum of the
cytoplasmic and nuclear receptors after the incubation of whole cells
with labelled glucocorticoids (Crabtree et al., 1978), an estimation
of the amount of the receptor present independently in the cytoplasm
and in the nucleus is obtained. This assay may result in some underestimation of cytoplasmic receptor by charcoal adsorption, particularly when low receptor levels are present.

A comparative analysis of both the whole cell assay and the
cytosol assay has recently been performed in our laboratory (Iacobelli et al., 1981 b). Since the amount of cells recovered from
most of the patients was not sufficient to furnish high enough cytosol
protein concentrations in the sample, we decided to use the DE 81
ion-exchange filter assay which accurately measures GR levels at
very low protein concentrations (Santi et al., 1973). The results
of this study demonstrated that the two GR assay methods give correlated values, independently of the type of leukemia studied. The
correlation is less evident for cells with low receptor content.
Moreover, the cytosol assay consistently leads to an underestimation of the number of binding sites as compared to the values obtained by whole cell assay. Probably there are several factors that
could explain the reasons for this underestimation. First, it is
essential to remember that with the whole cell assay, the total
quantity of receptor, cytoplasmic plus nuclear, is measured. Therefore, since at $37°$, in the presence of physiological concentrations

of hormone, about 80% of the receptor is located in the nucleus, as
has been demonstrated in the rat (Munck and Wira, 1971) and in human
thymocites(Iacobelli et al., 1980 b), it is obvious that cytosol assay
gives a noticeable underestimation. Second, possible breakage of lysosomes and release of proteolytic enzymes that partially inactivate
the cytoplasmic receptor must also be considered. Our data show that
the receptor binding capacity in the cytosol from certain types of
leukemic cells decreases rather rapidly, suggesting that substances
capable of inactivating the receptor molecule are active in these
cells. Although these substances (possibly enzymes) have not yet
been identified, it is interesting that Sloman and Bell (1979) have
shown that the cytosol of CML contains receptor-inactivating factor
(s), which could be activated during the homogenization procedure.
The presence of similar factors in leukemic myeloblasts of AML has
also been proposed by other authors (Crabtree et al., 1978). Third,
the GR levels found in some patients greatly exceed those found in
normal leukocytes of both lymphatic and myelomonocytic origin (Neifeld et al., 1977). This capacity of the leukemic cell to express
a greater quantity of receptor protein with respect to its normal
counterpart seems to be a more general characteristic of the neoplastic phenotype. It is possible to hypothesize that, in the presence of "redundant" GR in leukemic cells, the nuclear acceptor -
binding capacity could become the limiting step in the traslocation
process. Therefore, in cells with low receptor content, the quantity
of binding sites measured in the cytoplasm could represent a minimal part of the total cellular receptor sites, which in the presence
of endogenous hormones are preferentially localized in the nucleus.
This could explain the high frequency of negative GR determination
obtained by cytosol assay in leukemic cells with low receptor levels.
However, when the cellular receptor levels are very high, they exceed the nuclear acceptor capacity and a high proportion of binding
sites are found in the cytoplasmic compartment.

In brief, these results suggest that for the determination
of GR in leukemic cells it is preferable to use whole cell assay
which gives an estimation of the total receptor content because
the measurement of cytoplasmic binding sites alone may lead to
falsely - ve results. On the other hand, demonstration of total receptor sites in whole cell preparation does not give information
about the localization, i.e., if in the cytoplasm and/or in the
nucleus. While our conclusions may be used to partly explain the
contradictory data on GR determination in leukemic cells, at least
in cases of low receptor content, they indicate that receptor assess-
ments by whole cell and cytosol assays provide non equivalent but
possibly complementary information.

Glucocorticoid receptors and response to endocrine therapy

In leukemic cells, the in vitro effects of glucocorticoids

on DNA, RNA, and protein synthesis, cell viability and other biological parameters are highly variable and poorly related to either
clinical features of the disease, cellular levels of GR or in vivo
response to glucocorticoids (Kontula, 1980; Bell, 1982). Some correlations between in vitro actions of glucocorticoids and GR levels
have been observed, even if at a variable extent, in ALL
(Cline and Rosenbaum, 1968; Lippman et al., 1973 a; Crabtree et
al., 1978), in AML (Crabtree et al., 1981) and in non-Hodgkin's
lymphoma (Bloomfield et al., 1980). However, other investigators
obtained opposite results (Homo et al., 1978; Sloman and Bell,
1979; Homo et al., 1980; Kontula et al., 1980; Iacobelli et al.,
1980 b). Among the factors explaining these discrepancies are the
heterogeneity of the leukemic cell population and the artificial
subtraction of influences present in vivo, which may condition
the cell response to glucocorticoids (Iacobelli et al., 1978;
Homo et al., 1980).

   In an attempt to identify those patients likely to respond
to glucocorticoids, we have correlated the GR levels in leukemic
lymphoblasts with the response to short-term treatment with glucocorticoids alone in 19 children with ALL (Mastrangelo et al.,
1980). GR were present in all patients,even though the receptor
concentrations varied considerably from one patient to another
(range: 688-20,791 sites/cell). A significant difference in the
number of GR sites in leukemic cells between responsive and non
responsive patients was observed. A similar difference in GR concentrations between responsive and nonresponsive patients treated
with glucocorticoids alone, has recently been reported in cases of
adult ALL (Bloomfield et al., 1981) and in non-Hodgkin's lymphoma
(Bloomfield et al., 1980). On the other hand, Homo et al.(1980)
found no significant differences in leukocytes GR content between
glucocorticoids responsive and unresponsive patients. However,
only a few patients were investigated and no comparison was made
between the number of receptor sites/cell and the magnitude of
the response to corticosteroids for each single patient. The apparent absence of correlation may merely reflect the misleading
determination of the mean value for the two groups of patients.
The main clinical implication of these observations concerns the
usefulness of GR assay in the selection of those ALL patients with
lymphoblasts containing few GR, in whom only untoward reactions
of corticosteroids may be expected without beneficial effects.

   Studies on the clinical significance of GR in AML have fur-
nished partially conflicting results. In fact, Lippamn et al.,(1975)
found a relationship between the presence and the concentration
of GR and the response to glucocorticoid treatment both in vitro
and in vivo, while other authors (Crabtree et al., 1978; Iacobelli
et al., 1978; Crabtree et al., 1981) failed to find a similar
correlation.

Few data are actually available on the clinical relevance of GR determination in CLL and CML (Kontula, 1980; Bell, 1982).

Glucocorticoid receptors, response to chemotherapy and prognosis

In 1973 (b) Lippman et al. reported a good correlation between response to polychemotherapy and the presence of GR. Lymphoblasts of 22 patients studied, who were responsive to therapy, contained appreciable GR levels as did those of 6 patients in relapse who sub-sequentely responded to the same treatment. On the other hand, the cells from 6 other patients in relapse, who were resistant to the therapy, contained low GR levels. Subsequentely, Lippman et al. (1978 b) demonstrated a positive correlation between the levels of GR in lymphoblastsof patients with ALL and remission duration and survival independently of other factors such as age, sex, white cell blood count and presence of surface markers. Therefore the presence of GR was considered an important biochemical marker with prognostic significance, although this concept has been questioned by other investigators (Duval and Homo, 1978; Kontula et al., 1980; Bloomfield et al., 1981).

Using a whole cell assay we have measured GR levels in lympho-blasts from 78 previously untreated ALL patients and from 10 patients in relapse (Marchetti et al., 1982). 61 patients achieved complete remission following intensive combination chemotherapy; 4 patients died during the induction period. The GR concentrations found in patients who did (CR+) or did not (CR-) go into complete remission were significantly different (CR+: median, 10, 063 sites/cell; range 884-41, 560 sites/cell. CR-: median 3,078 sites/cell; range 312-9,351 sites/cell). Moreover, all of the patients studied in relapse showed low GR levels. The analysis of the actuarial curves of the disease-free interval and the survival showed a significantly more favourable course in the group of patients with lymphoblasts containing high GR levels (more than 6,000 sites/cell).

Whereas in our previous study (Mastrangelo et al., 1980) we observed that patients with lymphoblasts containing less than 4,000 sites/cell appeared resistant to a short-term treatment with predni-sone, we recently found two patients with low levels of GR dramatically responding to glucocorticoids alone. In order to acquire a better understanding of the GR significance in the management of all we investigated the relationship between GR response to glucocorticoid treatment and prognosis (Marchetti et al., 1983 a).25 children with newly diagnosed ALL were evaluated. Prednisone was given at 2 mg/kg/day for three days before combination polychemotherapy. Response to treatment was evaluated on the fourth day as the change in the total peripheral blast count. In 14 patients we observed a complete disappearance of peripheral blasts, whereas in the remaining 11

patients nearly no effect was produced. 3 non responder patients
did not achieve bone marrow remission. A trend toward a correlation
between response to glucocorticoids and prognosis was observed:
8/14 responders are in complete remission (1.5 to 5.0 years
follow up; median, 2.9 years). In contrast only 2/8 evaluable non-
responders are still in complete remission (7 months and 4.1 years,
respectively). Moreover, there was a significant difference in the
number of GR between those patients who did and did not respond to
treatment with glucocorticoids. Density of receptors was strictly
correlated with inducibility and duration of complete remission.
It is interesting to note that all responding patients actually
in complete remission showed high lymphoblast GR levels at the
time of the diagnosis . Even if further correlative clinical studies
are necessary, it seems reasonable to conclude that the response
to the in vivo test plus receptor assay represent a real progress
in the prediction of the clinical course of the disease.

While it is not yet known how GR work as prognostic indicators,
it is possible to hypothesized that the receptor density in the
neoplastic cell is connected to the degree of differentiation.
For example, in human thymus GR content of the lymphoid cells
increases as the cell approaches a higher degree of immunologic
maturity (Ranelletti et al.,1981).

The GR content in lymphatic cells is higher during the synthetic
phase of the cycle (Cidlowski and Michaels, 1977) and in cells
blastically transformed by mitogens (Neifeld et al, 1977). It is
therefore possible that the high GR levels found in the lymphoblasts
of patients sensitive to therapy are simply a reflection of a
higher percentage of cells in the synthetic phase of the cycle
and more sensitive to the cycle-specific drugs used. More knowledge
of these aspects are necessary for a more correct evaluation of
the complex relationship between glucocorticoids and cell responsiveness.

Markers for functioning receptor sites

The importance of measuring specific markers for functioning
GR in leukemia is quite evident. We have demonstrated (Marchetti et
al., 1983 b) that the enzyme glutamine-synthetase is present in the
lymphoblasts of patients with ALL and that its activity can be
specifically regulated by glucocorticoids. A direct correlation was
found between the magnitude of glucocorticoid mediated increase
of glutamine-synthetase activity and the cellular levels of GR.
In a few receptor-positive specimens no induction was observed,
despite the presence of the cytoplasm-to-nucleus translocation
of the receptor-steroid complex. The steroid-inducible glutamine-
synthetase in leukemia lymphoblasts may represent a useful indicator
of the integrity of the receptorial machinery.

REFERENCES

Adams,J.B. and Wong, S.F.,1969, Desmolase activity of normal and
    malignant human breast tissue, J.Endocrinol., 44:69.
Allegra,J.C., Barlock,A., Huff,K., and Lippman,M.E.,1980, Changes
    in multiple or sequential estrogen receptor determinations
    in breast cancer, Cancer, 45:792.
A.M.A., American Medical Association Committee on Research, 1960,
    Androgens and estrogens in the treatment of disseminated
    mammary carcinoma, J.Amer.Med.Assoc., 172:1271.
Anderson,D.G.,1972, The possible mechanism of action of progestins
    on endometrial adenocarcinoma, Amer.J. Obstet.Gynec., 113:
    195.
Anderson, W.A., Kang, Y.H., and De Sombre, E.R., 1975, Endogenous
    peroxidase: a specific marker enzyme for tissue displaying
    growth dependency on estrogens, J.Cell.Biol., 64:668.
Beatson,G.T.,1896, On the treatment of inoperable cases of carci-
    noma of the mamma. Suggestion for a new method of treatment
    with illustrative cases, Lancet, ii:104
Bell,P.A.,1982, Glucocorticoids in the therapy of leukemia and
    lymphoma, in: "Clinics in Oncology. Hormone Therapy", B.J.A.
    Furr, ed., W.B. Saunders Co.Ltd., London.
Bird,C.C.,1979, Clinical classification of leukemia and lymphoma
    in relation to glucocorticoid therapy, in: "Glucocorticoid
    action and leukemia", P.A.Bell, N.M.Borthwick, eds., Alpha
    Omega Publ.Ltd.,Cardiff.
Blamey,R.W., Bishop,H.M., Blake,J.R.S., Doyle,P.J., Elston,C.W.,
    Haybittle,J.L., Nicholson,R.I., Griffiths,K.,1980, Relation-
    ship between primary breast tumor receptor status and patient
    survival, Cancer, 46:2765.
Bloom,N.D., Tobin,E.H., and Schreibman,B.,1980, The role of proge-
    sterone receptor in the management of advanced breast cancer,
    Cancer, 45:2992.
Bloomfield,C.D., Smith,K.A., Peterson,B.A., Hildebrandt,L.,
    Zaleskas,J., Gajl-Peczalska,K.J., Frizzera,G., and Munck,A.,
    1980, In vitro glucocorticoid studies for predicting response
    to glucocorticoid therapy in adults with malignant lymphoma,
    Lancet, i:952.
Bloomfield,C.D., Kendall,A.S., Peterson,B.A., and Munk,A., 1981,
    Glucocorticoid receptors in adult acute Lymphoblastic leukemia.
    Cancer Res., 41:4857.
Bonadonna,G., Valagussa,P., Rossi,A., Tancini,G., Brambilla,C.,
    Marchini,S., and Veronesi,U., 1981, Multimodal therapy with
    CMF in resectable breast cancer with positive axillary nodes.
    The Milan Institute Experience, in :"Adjuvant therapy of
    cancer III" , S.E.Salmon and S.E.Jones,eds. Grune and Stratton
    Publ., New York.
Bonte,J.,1972, Medroxyprogesterone in the management of primary
    and recurrent metastatic adenocarcinoma, Acta Obstet.Gynec.

Scand., 51(Suppl.19):21.

Bonte,J.,1980, Hormonal dependence of endometrial adenocarcinoma and
        its hormonal sensitivity to progestogens and antiestrogens,
        in: "Hormones and Cancer," S.Iacobelli, R.J.B.King, H.R.
        Lindner, and M.E.Lippman eds., Raven Press, New York.

Bonte,J., Decoster, J.M., Ide, P., and Billiet, G.,1978, Hormono-
        prophylaxis and hormonotherapy in the treatment of endome-
        trial adenocarcinoma by means of medroxyprogesterone acetate.
        Gynec.Oncol.,6:60.

Bonte,J.,Ide, P., Billiet,G., Wynants,P., 1981, Tamoxifen as a pos-
        sible chemotherapeutic agent in endometrial adenocarcinoma,
        Gynec. Oncol., 11:140.

Braunsberg,H., James, V.H.T., Jamieson, C.W., Desai, S., Carter, A.E.
        and Hulbert, M., 1974, Effect of age and menopausal status
        on estimates of estrogen binding by human malignant breast
        tumours, Brit.Med.J., 791-745.

Broens, J., Mouridsen, M.T., and Soerensen, H.M.,1980, Tamoxifen
        in advanced endometrial carcinoma, Cancer Chemother.Pharma-
        col., 4:213.

Brooks, S.C., Saunders, D.E., Singhakowinta, A., and Vaitkevicius,
        V.K., 1980, Relation of tumor content of estrogen and pro-
        gesterone receptors with response of patient to endocrine
        therapy, Cancer, 46: 2775.

Bruckner,H.W. and DEPPE, G., 1977, Combination chemotherapy of ad-
        vanced endometrial adenocarcinoma with adriamycin, cyclo-
        phosphamide, 5-fluorouracil, and medroxyprogesterone aceta-
        te, Obstet. Gynec., 50:105.

Carbone, P.P., and Carter, S.K.,1974, Approach to development of
        effective chemotherapy, Gynec.Oncol., 2:348.

Chamness,G.C., and McGuire W.L., 1975, Scatchard plots: common
        errors in correction and interpretation, Steroids, 26:538.

Chamness, G.C;, and McGuire, W.L., 1979, Methods for analysing ste-
        roid receptors in human breast cancer, in: "Breast Cancer
        3.Advances in Research and Treatment," W.L.McGuire ed.,
        Plenum Press, New York.

Cidlowski, J.A., and Michaels, G.A., 1977, Alteration in glucocor-
        ticoid binding site number during the cell cycle in the HeLa
        cells, Nature, 266: 643.

Claman , N.H., 1972, Corticosteroids and Lymphoid cells, New Engl.
        J.Med., 287:388

Clark J.H., and Peck, E.J.Jr., 1979, Female sex steroids.Receptors
        and function. Monographs on Endocrinology, Springer Verlag,
        Berlin, Heidelberg, New York.

Cline, M.J., and Rosenbaum, E., 1980, Prediction of in vivo cyto-
        toxicity of chemotherapeutic agents by their in vitro effect
        on leukocytes from patients with acute leukaemia, Cancer Res.
        28: 2516.

Cohen, C.J., Deppe, G., and Bruckner, H.W., 1977, Treatment of ad-
        vanced adenocarcinoma of the endometrium with melphalan, 5-

fluorouracil, and medroxyprogesterone acetate, Obstet.Gynecol., 50: 415.

Cooper, A.P., 1835, Lectures on the principles and practice of surgery, John Thomas Cox, London

Crabtree, G.R., Smith, R.A., and Munck, A., 1978, Glucocorticoid receptors and sensitivity of isolated human leukemia and lymhoma cells, Cancer Res., 38: 4268.

Crabtree, G.R., Bloomfield, C.D., Smith, K.A., McKenna, R.W., Peterson, B.A., Hildebrandt, L., and Munck, A., 1981, Glucocorticoid receptors and in vitro responses to glucocorticoid in acute non lymphocytic leukemia, Cancer Res., 41: 4853.

Crocker, S.G., Milton, P.J.D., and King, R.J.B., 1974, Uptake of (6,7-^{3}H)oestradiol-17 by normal and abnormal human endometrium, J.Endocr., 62:145.

Daniel, L., Martin, P., Escrich, E., Tubiana, N., Fiere, D., and Saez, 1981, Androgen, estrogen and progestin binding sites in human leukemic cells, Int.J.Cancer, 27: 733.

Dao, T.L., and Nemoto, T.,1980, Steroid receptors and response to endocrine ablation in women with metastatic cancer of the breast, Cancer, 46: 2779.

Degenshein, G.A., Bloom, N., and Tobin, E., 1980, The value of progesterone assays in the management of advanced cancer, Cancer, 46:2789.

Deppe, G., Bruckner, M.W., and Cohen, C.J., 1980, Combination chemotherapy for advanced endometrial adenocarcinoma, Int. J. Gynaec.Obstet., 18:168.

De Sombre, E.R., 1982, 1982, Breast Cancer: hormone receptors, prognosis and therapy, in:"Clinics in Oncology.Hormone therapy," 1,1, B.J.A. Furr, ed., W.B.Saunders Co.L.t.d.,London.

De Sombre, E.R. and Jensen, E.V., 1980, Estrophilin assays in breast cancer: quantitative features and application to the mastectomy specimen, Cancer, 46: 2783.

Donovan, J.F., 1974, Nonhormonal chemotherapy for endometrial adenocarcinoma: a review. Cancer, 34: 1587.

Dougherty, T.F., 1952, Effect of hormones on lymphatic tissues, Physiol.Rev., 32: 379.

Dougherty, T.F., Berliner, M.L., Schneebeli, G.L., and Berliner, D.L., 1964, Hormonal control of lymphatic structure and function, Am.N.Y.Acad.Sci., 113:825.

Duval, D., and Homo, F., 1978, Prognostic value of steroid receptor determination in leukemia, Cancer Res., 38: 4263.

Ehrlich, C.E., and Young, P.C.M., 1980, Prediction of hormone-sensitivity in endometrial cancer, Rev.Endocrine-related cancer, 6: 13.

Feherty, P., Robertson, D.M., Waynforth, H.B., and Kellie, A.E.,1970, Changes in the concentrations of high affinity oestradiol receptors in rat uterine supernatant preparations during the oestrus cycle, pseudopregnancy, pregnancy, maturation and after ovariectomy, Biochem.J., 120: 837

Fidler,I.J. and Hardt,I.R., 1981, The origin of metastatic heteroge-
     neity in tumors. Europ.J.Cancer,17:487.
Folca,P.J., Glascock,R.F. and Irvine,W.T., 1961, Studies with tritium
     labelled hexoestrol in advanced breast cancer, Lancet, 2:796.
Furr, B.J.A., ed., 1982, "Clinics in Oncology. Hormone Therapy,"
     W.B. Saunders Co. Ltd., London.
Gailani,S., Minowada,J., Silvermail, P., Nussbaum,A., Kalser,N.,
     Rosen,F., and Shimaoka,K.,1975, Specific glucocorticoid binding
     in human hemopoietic cell lines and neoplastic tissues, Cancer Re
     33:2653.
Gehring,U., 1980, Genetic and biochemical studies on glucocorticoid
     receptors, in:" Hormones and Cancer", S.Iacobelli, R.J.B.King,
     H.R.Lindner, and M.E.Lippman, eds., Raven Press, New York.
Goldin,A., Sandberg,J., Henderson,E.,Newman,J., Frei,E., and Holland,
     J., 1971, The chemotherapy of human and animal acute leukemia,
     Cancer Chemother. Rept., 55:309
Griffiths,K., Maynard,P.V., Wilson,D.W., and Davies,P., 1978, Endo-
     crine aspects of primary breast cancer, in:"Review on Endocrine-
     related Cancer, "M.Mayer, S.Saer, and B.A.Stoll, eds., I.C.I.
     Publishing,U.K.
Gurpide,E.,1981, Hormone receptors in endometrial cancer, Cancer,
     48:638.
Hahnel,R.and Vivian,A.B., 1975, Biochemical and clinical experience
     with the estimation of estrogen receptors in human breast carci-
     noma, in: "Estrogen Receptors in Human Breast Cancer", W.L. Mc
     Guire, P.P.Carbone, and E.P.Vollmer,eds., Raven Press, New York.
Hamburger,A.W., Salmon,S.E., and Soehnlen,B.,1978, Quantitation of
     differential sensitivity of human-tumor stem cells to anticancer
     drugs, N.Engl.J.Med., 298:1321.
Heuson,J.C., Leclercq,G., Longeval,E.,Deboel,M.C.,Mattheiem,W.H.,
     and Heiman,R.,1975, Estrogen receptors: Prognostic significance
     in breast cancer, in: "Estrogen Receptors in Human Breast Cancer"
     W.L.Mc Guire, P.P. Carbone and E.P.Vollmer,eds., Raven Press
     New York.
Homo, F., Duval,D., and Meyer,P., 1975, Etude de la liason de la
     dexamethasone tritiee dans le lymphocytes des sujets normaux
     et leucemiques, C.R.Acad.Sci.Paris, t 280: D 1923.
Homo,F., Duval,D.,Meyer,P.,Debre,P., and Binet,J.L., 1978,Cronic
     lymphatic leukemia: cellular effect of glucocorticoids in vitro,
     Br.J.Haematol., 38:491.
Homo,F.,Duval,D.,Harousseau,J.L., Marie,J.P., and Zittoun,R., 1980,
     Heterogeneity of the in vitro responses to glucocorticoids in
     acute leukemia, Cancer Res., 40:2601.
Horwitz,K.B., and Mc guire,W.L., 1975, Specific progesterone receptors
     in human breast cancer, Steroids, 25:497.
Horwitz,K.B., and Mc Guire,W.L., 1978, Estrogen control of progeste-
     rone receptor in human breast cancer. Correlation with nuclear
     processing of estrogen receptors, J.Biol.Chem., 253:2223.
Horwitz,K.B., Mc Guire, W.L., Pearson, O.H., and Segaloff, A., 1975

Predicting response to endocrine therapy in human breast cancer: an hypothesis, Science, 189:726.

Horwitz, K.B., Koseki,Y, and Mc Guire, W.L., 1978, Estrogen control of progesterone receptors in experimental breast cancer: role of estradiol and antiestrogen, Endocrinology, 103:1742.

Hsueh,A.J.W., Peck,E.J.,and Clark,J.H., 1976, Control of uterine estrogen receptor levels by progesterone, Endocrinology, 98:438.

Iacobelli,S., Ranelletti,F.O., Longo,P., Riccardi,R., and Mastrangelo, R., 1978, Discrepancies between in vivo and in vitro effects of glucocorticoids in myelomonocytic leukemia cells with steroid receptors, Cancer Res., 38:4257.

Iacobelli,S., Longo,P., Scambia,G., Natoli,V., and Sacco,F., 1980a, Progesterone receptors and hormone sensitivity of human endometrial carcinoma, in: "Role of Medroxyprogesterone in Endocrine-Related Tumors," S.Iacobelli, A.Di Marco, eds., Raven Press New York.

Iacobelli,S., Longo,P., Mastrangelo, R., Malandrino,R., Ranelletti, F.O., 1980b, Glucocorticoid receptors and steroid sensitivity of acute lymphoblastic leukemia and thymoma, in:"Hormones and Cancer," S.Iacobelli, R.J.B. King, H.Lindner, M.E.Lippman, eds., Raven Press, New York.

Iacobelli, S., King, R.J.B., Lindner, H.R., and Lippman, M.E., 1980c, Hormones and Cancer, Raven Press, New York.

Iacobelli,S., Marchetti,P., Bartoccioni,E.,Natoli,V., Scambia,G., and Kaye,A.M.,1981a, Steroid-induced proteins in human endometrium, Molec.Cell Endocrinol., 23:321.

Iacobelli,S., Natoli,V., Longo,P., Ranelletti,F.O., De Rossi,G., Pasqualetti,D., Mandelli,F., and Mastrangelo,R., 1981b, Glucocorticoid receptor determinations in leukemia patients using cytosol and whole cell-assay, Cancer Res. 41: 3979.

Iacobelli,S., Longo,P., Natoli,V., Sica,G., Bartoccioni,E., Marchetti, P., and Scambia G., 1982, New tools of potential value for predicting hormone responsiveness in human endometrial cancer, in: "The Menopause: Clinical, Endocrinological, and Pathophysiological Aspects," P.Fioretti, L.Martini, G.B.Melis, S.S.C. Yen, eds., Academic Press, New York.

Iacobelli,S., Natoli,V., Natoli,C., Marchetti,P., Scambia, G., and Kaye, A.M., 1983a, Estrogen induced proteins in normal and neoplastic tissues, in: "Proceedings of the Second Innsbruck Winter Conference on Biochemistry in Clinical Medicine. Hormone-cell interactions in reproductive tissue" (in press), Masson Publ. Inc.New York.

Iacobelli, S., Sica, G., Marchetti, P., Natoli, C., and Natoli, V., 1983 b, Ovarian steroids and tumors, in : "The Ovary," G.B.Serra, ed., Raven Press, New York.

Isomaa, V., Isotalo, H., Orava, M., and Janne, O., 1979, Regulation of cytosol and nuclear progesterone receptors in rabbit uterus by estrogen, antiestrogen and progesterone administration,

Biochim.Biophys.Acta, 585:24.

Janne,O.,Kauppila, A., Kontula, K.,Syrjala, P.,and Vihko, R., 1977,
    Female sex steroid receptors in normal, hyperplastic and car-
    cinomatous endometrium. The relationship to serum ste-
    roid hormones and gonadotropins and changes during medroxy-
    progesterone acetate administration, Int.J.Cancer, 24 :545.

Jensen, E.V., and Jacobson, H.I., 1962, Basis guides to the mechanism
    of estrogen action. Rec. Prog. Horm. Res., 18: 387.

Jensen, E.V., Polley, T.Z., Smith, S., Block, G.E., Ferguson, D.J.
    and De Sombre, E.R., 1975a, Prediction of hormone dependency
    in human breast cancer, in: "Estrogen Receptors in Human Breast
    Cancer," W.L.Mc Guire, P.P.Carbone, and E.P.Vollmer, eds.,
    Raven Press, New York.

Jensen, E.V., Smith, S., Moran, E.M., and De Sombre, E.R., 1975b,
    Estrogen receptors and hormone dependency in human breast can-
    cers, in: "Hormones and Breast Cancer," M. Namer, ed., INSERM,
    Paris.

Jonat, W., Maass, H., Stolzenbach, G., and Trams, G., 1980, Estrogen
    receptor status and response to polychemotherapy in advanced
    breast cancer, Cancer, 46: 2809.

Kaye, A.M., Reiss, N., Shaer, A., Sluyser, M., Iacobelli, S., Amroch,
    D., and Soffer, Y., 1981, Estrogen responsive creatin-kinase
    in normal and neoplastic cells, J. Steroid Biochem., 15: 69.

Kelley, R.M. and Baker, W.H., 1961, Progestational agents in the
    treatment of carcinoma of the endometrium, New England J. Med.
    264: 216.

King, R.J., 1980, Analysis of estradiol and progesterone receptors
    in early and advanced breast tumors, Cancer, 46: 2818.

Kjellestrand, C.M., 1975, Side effect of steroids and their treatment
    Transpl. Proc., 7:123.

Knight III,W.A., Livingston,R.B., Gregory,E.J.,and MC Guire,W.L.,
    1977, Estrogen receptor is an independent prognostic factor
    for early recurrence in breast cancer, Cancer Res., 37:4669.

Knight III,W.A., Clark,G.M., Osborne,C.K., and Mc Guire, W.L., 1981,
    Intensive chemotherapy for estrogen receptor negative, stage
    II breast cancer, in: "Adjuvant therapy of cancer III," S.E.
    Salmon and S.E.Jones, eds., Grune and Stratton Publ., New
    York.

Kohorn,E.I., 1976, Gestagens and endometrial carcinoma, Gynec.Oncol.,
    4:398.

Kontula,K., 1980, Glucocorticoid receptors and their role in human
    disease, Ann.Clin.Res., 12:233.

Kontula,K., Andersson,L.C., Paavonen,P., Myllyla,G., Teerenhovi,L.,
    and Vuopio, P., 1980, Glucocorticoid receptors and glucocortico
    sensitivity of human leukemia cells, Inter.J.Cancer, 26:177.

Kute,T.E., Muss,H.B., Anderson,D., and Wittliff,J.L., 1981,Relation-
    ship of steroid receptor and cell kinetics and clinical status
    in patients with breast cancer. Cancer Res., 41:3524.

Leake,R.E., Laing,L., Calmon,K.C., 1981, Oestrogen-receptor status stability, Br. J. Cancer, 43:59.

Leclercq, G. and Heuson, J.C., 1977, Therapeutic significance of sex
    steroid hormone receptors in the treatment of breast cancer,
    Europ.J.Cancer, 13:1205.

Leclercq,G., Heuson,J.C., Deboel,M.C., and Mattheiem,W.H., 1975,
    Oestrogen receptors in breast cancer: a changing concept, Br.
    Med.J., i:185.

Lerner, L.J., 1964, Hormone antagonists: inhibitors of specific ac-
    tivities of estrogen and androgen, Recent Prog.Horm.Res., 20:
    435.

Lewis,G.C.Jr., Slack,N.H., Martel, R., and Bross,I.D.J., 1974,
    Adjuvant primary definitive treatment of endometrial cancer,
    Gynec.Oncol., 2:368.

Lippman,M.E.,1977, Steroid hormone receptors in human malignancy,
    Life Sci., 18:143.

Lippman,M.E., 1980, Steroid receptor analysis and endocrine therapy
    of breast cancer, in: "Perspectives in steroid receptor research",
    F.Bresciani, ed., Raven Press, New York.

Lippman,M.E., and Allegra,J.C., 1978, Estrogen receptor and endocrine
    therapy of breast cancer, N.Engl.J.Med., 299:930.

Lippman,M.E., and Allegra,J.C., 1980, Quantitative estrogen receptors
    analysis: the response to endocrine cytotoxic chemotherapy in
    human breast cancer and the disease free interval, Cancer,
    46:2829.

Lippman,M.E., and Nawata,H., 1982, A genetic analysis of estrogen
    action in human breast cancer cells, in: " The role of tamoxifen
    in breast cancer," S.Iacobelli, M.E.Lippman, and G.Robustelli
    Della Cuna, eds., Raven Press, New York.

Lippman,M.E., and Thompson,E.B., 1979, Pitfalls in the interpretation
    of steroid receptor assays in clinical medicine, in: "Steroid
    receptors and the management of cancer," vol.1, E.B.Thompson
    and M.E.Lippman, eds., Boca Raton, Florida, CRC Press.

Lippman, M.E., Halterman,R.H., Leventhal,B.G., Perry,S., and Thompson
    E.B., 1973a, Glucocorticoid-binding proteins in acute lympho-
    blastic leukemia blast cells, J. Clin. Invest., 72: 1715.

Lippman , M.E., Halterman, R.H., Perry,S., Leventhal,B.G., Thompson,
    E.B., 1973b, Glucocorticoid binding proteins in human leukemic
    lymphoblasts, Nature (New Biol.), 242:157.

Lippman, M.E., Perry,S., and Thompson, E.B., 1975, Glucocorticoid
    binding proteins in myeloblast of acute myelogenous leukemia,
    Am. J. Med., 59:224.

Lippman, M.E., Huff, K.K., and Bolan, G., 1977, Progesterone and
    glucocorticoid interactions with receptors in breast cancer
    cells in long term culture, N.Y.A.S., 286:101.

Lippman, M.E., Allegra, J.C., Thompson, E.B., 1978a, The relation
    between estrogen receptors and response rate to cytotoxic
    chemotherapy in metastatic breast cancer, N.Engl.J.Med.,298:
    1223.

Lippman,M.E., Yarbro,G.K.,and Leventhal,B.G., 1978b, Clinical

implications of glucocorticoid receptors in human leukemia,
Cancer Res., 38:4251
Livingstone,R.B., and Carter,S.K., eds., 1970, Single agents in
cancer chemotherapy, Plenum Press, New York.
Lubbert,H., and Pollow,K., 1978, Correlation between the 17 hydroxysteroid dehydrogenase activity and the estradiol and progeste-
rone receptor concentrations of normal and neoplastic mammary
tissue, J.Molecular Med., ':175.
Maas, H., Engel, B., Nowakowski, H., 1975, Estrogen receptors in human
breast cancer and clinical correlations, in:" Estrogen Receptors
in Human Breast Cancer," W.L.Mc Guire, P.P.Carbone, and E.P.
Vollmer, eds., Raven Press, New York.
Mac Laughlin, D.T. and Richardson, G.S., 1976, Progesterone binding
by normal and abnormal human endometrium, J. Clin. Endocrinol.
Metab., 42:667.
Malkasian, G.D., Decker, D.G., Mossey, E., Johnson, C.E. 1971, Pro-
gestin treatment of recurrent endometrial carcinoma, Am. J.
Obstet. Gynecol., 110:15.
Mandelli, F., Amadori, S., Cantù Rajnaldi,A., Cordero di Montezemolo,
L., Madon,E., Masera,G., Meloni,G., Pacilli,L., Paolucci,G.,
Pastore,G., Rosito,P., Uderzo,C., and Vecchi,V., 1980, Discon-
tinuing therapy in childood acute lymphocitic leukemia.A multi-
centric survey in Italy, Cancer, 46:1319.
Manni,A., Arafah,B., and Pearson,O.H., 1980, Estrogen and progesterone
receptors in the prediction of response of breast cancer to
endocrine therapy, Cancer, 46:2838.
Marchetti,P;, Natoli,V., Ranelletti,F.O., Mandelli,F., De Rossi,G.,
and Iacobelli,S., 1981, Glucocorticoids receptors studies in
human leukemia, J.Steroid Biochem., 15:261.
Marchetti,P., Spina,M.A., De Rossi,G., Guglielmi,C., Lopez,M., and
Iacobelli,S., 1982, Prognostic value of glucocorticoid receptors
in acute lymphoblastic leukemia, Proceedings of 3rd International
Symposium on Therapy of Acute Leukemias, Rome.
Marchetti,P., Riccardi,R., Mastrangelo,R., and Iacobelli,S., 1983a,
Glucocorticoid receptors and "in vivo" corticosensitivity of
leukemia lymphoblasts, in: Proceedings of 13th International
Congress of Chemotherapy, Vienna.
Marchetti,P., Ranelletti,F.O., Natoli,V., Sica,G., De Rossi,G., Iaco-
belli,S., 1983b, Presence and steroid inducibility of glutamine
synthetase in human leukemic cells, J.Steroid Biochem., in press
Martin,P.M., Rolland,P.H., Gammerre,M., Serment,H., and Toga,M., 1979
Estradiol and progesterone receptors in normal and neoplastic
endometrium: correlations between receptors,histopathologic
examination and clinical response under progestin therapy, Int.
J.Cancer, 23:321.
Mastrangelo,R., Malandrino,R., Riccardi,R., Longo,P., Ranelletti,F.O.
and Iacobelli,S., 1980, Clinical implications of glucocorticoid
receptor studies in childood acute lymphoblastic leukemia, Blood
56:1036.
Matsumoto,K., Ochi,H., Nomura,Y., 1978, Progesterone and estrogen

receptors in Japanese breast cancer, in: "Hormone receptors
and breast cancer," W.L. Mc Guire, ed., Raven Press, New York.

Maynard,P.V., Blamey,R.W., Elston,C.W., Haybittle,J.L., and Griffiths,
K., 1978, Estrogen receptor assay in primary breast cancer
and early recurrence of the disease, Cancer Res.,38:4292.

Mc Carty,K.S.Jr., Cox,C., Silva,J.S., Woodard,B.H., Mossler,J.A.,
Haagensen,D.E., Barton, T.K., Mc Carty,K.S.Sr., and Wells,S.A.,
1980, Comparison of sex steroid receptor analises and carcino-
embryonic antigen with clinical response to hormone therapy,
Cancer, 46:2846.

Mc Guire,W.L., 1980, The usefulness of steroid hormone receptors in
the management of primary and advanced breast cancer, in:
"Breast cancer: experimental and clinical aspects," H.T.
Mouridsen and Palshof, eds., Pergamon Press, New York.

Mc Guire,W.L., and Horwitz,K.B., 1978, Progesterone receptors in
breast cancer, in: "Hormone Receptors and Breast Cancer," W.L.
Mc Guire, ed., Raven Press, New York.

Mc Guire,W.L., Carbone,P.P., Sears,M.E., and Escher,G.C., 1975,
Estrogen receptors in human breast cancer: an overview, in:
"Estrogen receptors in human breast cancer", W.L.Mc Guire, P.P.
Carbone and E.P.Vollmer, eds., Raven Press, New York.

Mc Guire,W.L., Carbone,P.P., and Vollmer,E.P., eds., 1975b, "Estrogen
Receptors in Human Breast Cancer," Raven Press, New York.

Meyer,J.S., Rao, B.R., Stevens,S.C., and White,W.L., 1977, Low
incidence of estrogen receptors in breast carcinomas with rapid
rates of cellular replication, Cancer, 40:2290.

Milgrom,E, Thi,L., Atger,M., and Baulieu,E.E., 1973, Mechanism
regulating the concentration and the conformation of progesterone receptor(s) in the uterus, J.Biol.Chem., 248:6366.

Moon,H., 1936, Inhibition of somatic growth in castrate rats with
pituitary extract, Proc.Soc.Exptl.Biol.Med., 37:34.

Munck,A., and Vira,C., 1971, Glucocorticoid receptors in rat thymus
cells, in: "Advances in the Biosciences", G.L.Raspé, ed.,
Pergamon Press, Oxford, New York.

Namer, M., Lalanne,C., and Baulieu,E.E., 1980, Increase of progesterone receptor by tamoxifen as a hormonal challenge test in breast
cancer, Cancer Res., 40:1750.

Natoli,V., Marchetti,P., Natoli,C., Sica,G., and Iacobelli,S., 1981,
Estrogen and antiestrogen effects on protein synthesis in human
breast cancer cells, Curr.Chemother.Immunother., Suppl.1:1464.

Natoli,C., Sica,G., Natoli,V., Serra,A., and Iacobelli,S., 1983,
Two new estrogen-supersensitive variants of MCF-7 human breast
cancer cell line. Breast Cancer Res. and Treat., 3:1.

Nawata,H., Bronzert,D., and Lippman,M.E., 1981a, Isolation and
characterization of a tamoxifen-resistant cell-line derived from
MCF-7 human breast cancer cells, J.Biol.Chem., 256:5016.

Nawata,H., Chong,M.T., Bronzert,D., and Lippman,M.E., 1981b, Estradiol-
independent growth of a subline of MCF-7 human breast cancer cells
in culture, J.Biol.Chem., 256:6895.

Nenci,I, Beccati,M.D., Piffanelli,A., and Lanza,G., 1976, Detection
    and dynamic localization of estradiol-receptor complexes in
    intact target cells by immunofluorescence tecniques, J.Steroid
    Biochem., 7:505.
Neifeld,J.P., Lippman,M.E., and Tormey,D.C., 1977, Steroid hormone
    receptors in normal human lymphocytes. Induction of glucocorti-
    coid receptor activity by phytoemagglutinin stimulation, J.Biol.
    Chem., 252:2972.
Nicholson,R.I., Campbell,F.C., Blamey,R.W., Elston,C.W., George,D.,
    and Griffiths,K., 1981, Steroid receptors in early breast cancer:
    value in prognosis, J.Steroid Biochem., 15:193.
Nicole,I., and Saez,S., 1978, Human breast cancer isolated cells in
    short term culture: relation between steroid receptors content
    and the response to hormone addition, Cancer Res., 38:4314.
Noel,G., Becquart,D., and Maisin,H., 1981, Les recepteurs d'estroge-
    nes dans le cancer du sein. Effet de la radiotherapie pre-opera-
    toire, Senologia, 6:163.
Nomura,Y., Kobayashi,S., Takatani,I., Sugano,H., Matsumoto,D., and
    Mc Guire, W.L., 1977, Estrogen receptor and endocrine responsi-
    veness in Japanese versus American breast cancer patients,
    Cancer Res., 37:106.
Nomura,Y, Yamagata,J., Takenaka,K., and Tashiro,H., 1980, Steroid
    hormone receptors and clinical usefulness in human breast cancer,
    Cancer, 46:2880.
Nordqvist,R.S.B., 1964, Hormone effects on carcinoma of the human
    uterine body studied in organ culture. A preliminary report,
    Acta Obstet.Gynec.Scand., 43:296.
Nordqvist,R.S.B., 1970, Survival and hormonal responsiveness of endo-
    metrial carcinoma in organ culture, Acta Obstet.Gynec.Scand.,
    49:275.
Nordqvist,S. 1973, Endometrial cancer. Facts and theories, Path.Ann.,
    8:283.
Osborne, C.K., Yochmowitz,M.G., Knight,W.A.III, and Mc Guire,W.L.
    1980, The value of estrogen and progesterone receptors in the
    treatment of breast cancer, Cancer, 46:2884.
Panko,W.B., and Mac Leod, R.M., 1978, Unchanged nuclear receptors
    for estrogen in breast cancer, Cancer Res., 38:1948.
Paridaens,R., Sylvester,R.J., Ferrazzi,E., Legros,N., Leclercq,G.
    and Heuson,J.C., 1980, Clinical significance of the quantitative
    assessment of estrogen receptors in advanced breast cancer, Cance
    46:2889.
Piver,M.S., Barlow, J.J., Lurain,J.R., Blumenson,L.E., 1980, Medroxy-
    progesterone acetate (Depoprovera) vs hydroxyprogesterone
    caproate (Delalutin) in women with metastatic endometrial adeno-
    carcinoma, Cancer, 45:268.
Pollow,K., Lubbert,H., Boquoi,E., Kreuzer,G., and Pollow,B., 1975,
    Characterization and comparison of receptors for 17 estradiol
    and progesterone in human proliferative endometrium and endo-
    metrial carcinoma, Endocrinology, 96:319.
Pollow,K., Sinnecker,R., Schmidt-Gollwitzer,M, Boquoi,E., and Pollow,B

1977 a, Binding of ^{3}H-progesterone to normal and neoplastic tissue samples from tumor bearing breasts, J.Mol.Med. 2:60.

Pollow, K., Schmidt-Gollwitzer, M., and Nevinny-Stickel, J., 1977b, Progesterone receptors in normal human endometrium and endometrial carcinoma, in : " Progesterone receptors in normal and neoplasic tissues, " W.L.McGuire, J.P.Raynaud, E.E.Baulieu, eds., Raven Press, New York.

Pollow, K., Schmidt-Gollwitzer, M., and Pollow, B.,1980, Progesterone and estradiol binding proteins from normal human endometrium and endometrial carcinoma: a comparative study, in:"Steroid receptors and hormone-dependent neoplasia,"J.L.Wittliff, O.Dapunt, eds, Masson Publ., New York.

Poulsen,H.S., 1981, Oestrogen receptor assay limitation of the method. Europ.J.Cancer, 17:496

Ranelletti, F.O., Piantelli, M., Iacobelli, S., Musiani, P., Longo, P.,Lauriola, L., Marchetti, P;, 1981, Glucocorticoid receptors and in vitro corticosensitivity of peanut-positive and peanut-negative human thymocyte subpopulations, J. Immunol.,127:849

Richardson, G.S.,and Mac Laughin, D.T. eds., 1978, Hormonal Biology of Endometrial Cancer, vol.42, UICC Technical report series, Geneva.

Robel, P., Mortel, R., Levy, C., Namer, M., and Baulieu, E.E., Steroid receptors and response to an antiestrogen in post-menopausal endometrial carcinoma and metastatic breast cancer, in: "Non-steroidal antiestrogens molecular pharmacology and antitumor activity," R.L.Sutherland, ed. Academic Press, New York.

Rochefort, H., Pujol, H., Bressot, N., Lavie, M., Saussol, J., Verone, J., 1980, Effect de la radiotherapie sur la concentration des recepteurs d'estrogenes (RE) et progesterone (RP) dans le cancer du sein, in:" Recepteurs hormonaux et pathologie mammaire," P. M.Martin ed., Medsi Edit.Paris.

Rosen, P.T., Menendez-Botet, C.J., Nisselbaum J.S., Urban, J.A., Mike, V., Fracchia, A., and Schwartz, M.K., 1975, Pathological review of breast lesions analyzed for estrogen receptor protein, Cancer Res. 35: 3187.

Rozier, J.C., and Underwood, P.B., 1974, Use of progestational agents in endometrial adenocarcinoma, Obstet.Gynecol. 44:60.

Saez, S., Martin, P.M., and Chouvet, C.D., 1978, Estradiol and progesterone receptor levels in relation to plasma estrogen and progesterone levels, Cancer Res. 38: 3468.

Sakai, F., and Saez, S., 1976, Existence of receptors bound to endogenous estradiol in breast cancers of premenopausal and postmenopausal women, Steroids 27: 99.

Santi, D.,V., Sibley, C.H., Perriard, E.R., Thomkins, G.M., and Baxter, J.D., 1973, A filter assay for steroid hormone receptors,Biochemistry, 12:2412.

Schinzinger, A., 1889, Uber carcinoma mammae, Verhandlungen der Deutschen Gesellschaft fur Chirurgie, 18:28.

Schmidt, T.J., and Thompson, E.B., 1979, Glucocorticoid receptors and glutamine synthetase in leukemic Sezary cells, Cancer Res., 39:376

Seibert, K., and Lippman, M.E., 1982, Hormone receptors in breast
     cancer in: "Clinics in Oncology," 1,3, M.Baum, ed. W.B.Saunders,
     L.t.d., London.
Selye, H., 1937, Studies on adaptation, Endocrinology, 21:169.
Shaikh, B.S., Dougherty, J.B., Hamilton, R.W., Ballard, J.O., Patel,
     S.B., Gerviz, N.R., and Eyster, M.E., 1980, Selective use of dau-
     norubicine for remission-induction chemotherapy in acute non lym-
     phoblastic leukemia, Cancer, 46: 1731.
Sherman, A.I., 1966, Progesterone caproate in the treatment of endome-
     trial cancer, Obstet.Gynec., 28: 309.
Silvestrini, R., Daidone, M.G., and Di Fronzo, G., 1979, Relationship
     between proliferative activity and estrogen receptors in breast
     cancer, Cancer, 44: 665.
Singhakowinta, A., Mohindra, R., Brooks, S.C.,1975, Clinical correlati
     of endocrine therapy and estrogen receptor,in : " Estrogen Recep-
     tors in Human Breast Cancer," W.L.McGuire, P.P.Carbone, and E.P.
     Vollmer, eds., Raven Press, New York.
Skinner, L.G., Barnes, D.M., and Ribeiro, G.G., 1980, Simoultaneous e-
     stimation of cytoplasmic estrogen and progesterone receptors and
     nuclear estrogen receptors in human breast tumors and correlation
     with clinical response, in : " Steroid receptors and hormone de-
     pendent neoplasia, " J.L.Wittliff and O.Dapunt, eds., Masson Publ
     New York.
Sloman, J.C., and Bell, P.A., 1979, Glucocorticoids and myeloid leukem
     in :" Glucocorticoid action and leukemia,", P.A.Bell, N.M.Borthwi
     eds. Alpha Omega Publ. L.t.d., Cardiff.
Smith, J.P., 1978, Chemotherapy in gynecologic cancer, Surgical Clini
     of North Am, 58: 201.
Smith, R.G., 1980, Quality control in steroid hormone receptor assays,
     Cancer , 46: 2946.
Spiers, A.S.D., Goldman, B.M., Catowsky, D., Castello, C., Buskard, N.
     and Galton, D.A.G., 1977, Multiple drug chemotherapy for acute le
     kemia, Cancer , 40: 20.
Stevens, J., Stevens, Y.W., Sloan, E., Rosenthal, R., Rhodes, J., 1978
     Nuclear glucocorticoid binding in chronic lymphocytic leukemia ly
     phocytes, Endocrinol.Res.Comm.,5:91.
Stoll, B.A., 1969, "Hormonal management in breast cancer,", Pitman Med
     Publishing Company, L.t.d., London.
Stoll, B.A.ed., 1972,"Endocrine Therapy in Malignant Diseases," W.B.Sa
     ders Company L.,t.d., London.
Stoll, B.A., 1973, Hormonal therapy of gynecological malignancy, Clin.
     Obste. Gynec., 16:130.
Swenerton, K.D., 1980, Treatment of advanced endometrial adenocarcinom
     with tamoxifen, Cancer Treat.Rep, 64: 805.
Swenerton,K.D., 1982, Endometrial adenocarcinoma, in: "Clinics in onco
     logy. Hormone therapy," 1,1, B.J.A. Furr, ed., W.B.Saunders Co.
     L.t.d., London.
Takamirawa,H., and Sekiya,S., 1978, Steroid hormone action in uterine
     cancer, in: "Endocrine control in neoplasia," R.K.Sharma and W.E.
     Criss, eds., Raven Press, New York.

Terenius,L., Lindell,A., and Person,B.H., 1971, Binding of estradíol-
17 to human cancer tissue of the female genital tract, Cancer
Res. , 31:1895.
Terenius, L., Simonssons, B., and Nilssons, K., 1976, Glucocorticoid
receptor, DNA synthesis, membrane antigens and the relation to
disease activity in chronic lymphatic leukemia, J.Steroid.Bio-
chem., 7:905
Tseng, L., Gurpide, E., 1975 a, Effects of progestins on estradiol
receptor levels in human endometrium, J.Clin.Endocrinol.Metab.
41:402.
Tseng, L., Gurpide, E., 1975 b, Induction of human endometrial estradiol dehydrogenase by progestins,Endocrinology, 97: 825.
Vaughan, W.P., Karp, J.E., and Burke, P.J., 1980, Long chemotherapy-
free remissions after simple-cycle timed-sequential chemotherapy for acute myelocytic leukemia, Cancer, 45: 859.
Vietti, T.J., Sullivan, M.P., Berry, D.H., Hardy, T., Haggard, M. and
Blattner, R., 1965, The response of acute childhood leukemia to
an initial and a second course of prednisolone, J.Pediatr, 66:18.
Westley,B., and Rochefort,H., 1979, Estradiol induced proteins in the
MCF-7 human breast cancer cell line, Biochem. Biophys. Res. Comm.,
90:410.
Westley,B., and Rochefort,H., 1980, A secreted glycoprotein induced
by estrogen in human breast cancer cell line, Cell, 20:352.
Wittliff,J.L., 1980, Steroid receptor interactions in human breast
carcinoma, Cancer, 46:2953.
Wittliff,J.L., and Dapunt, O., eds., 1980, "Steroid receptors and
hormone-dependent neoplasia," Masson Publ., New York.
Wittliff,J.L., and Savlov,E.D., 1975, Estrogen-binding capacity of
cytoplasmic forms of the estrogen receptors in human breast
cancer, in: "Estrogen Receptors in Human Breast Cancer," W.L.
Mc Guire, P.P.Carbone, and E.P.Vollmer, eds., Raven Press, New
York.
Wittliff, J.L., Hilf, R., Brooks, W.F.Jr., Savlov, E.D., Hall, T.C.,
Orlando, R.A.,1971, Specific estrogen-binding capacity of the
cytoplasmic receptor in normal and neoplastic breast tissues
of humans, Cancer Res.32:1983.
Wittliff, J.L., Mehta, R.G., Lewko, W.M., and Park, D.C., 1982, Ste-
roid-receptor interactions in normal and neoplastic mammary
tissues, in: "Biochemical markers for cancer," T.Ming Chu, ed.,
M.Dekker Inc., New York.
Yamamoto, K.R., and Alberts, B.M., 1976, Steroid receptors: elements
for modulation of eukaryotic transcription, Ann.Rev.Biochem.,
45:721.
Young, P.C., 1979, Gynecologic malignancies, in:" Cancer Chemotherapy,"
H.M. Pinedo, ed., Elsevier, New York.
Young, P.C., Ehrlich, C.E., and Cleary, P.E.,1976, Progesterone binding
in human endometrial carcinomas, Am.J.Obstet.Gynecol., 125:353.
Young, P.C., Ehrlich, C.E., and Einhorn, L.H., 1980, Relationship
between steroid receptors and response to endocrine therapy and
cytotoxic chemotherapy in metastatic breast cancer, Cancer, 46:2961

THE ROLE OF PHOSPHOLIPIDS IN RECEPTOR BINDING IN THE NERVOUS SYSTEM

Michael Giesing[*]

Labor für Nervengewebekultur; Institut für
Physiologische Chemie der Universität
Nussallee 11; D-5300 Bonn;   G.F.R.

INTRODUCTION

Nerve tissue is richer in phospholipids (PL) than
most other parenchymal organs. Furthermore, neurons
and glia contain very small amounts of neutral lipids,
i.e. free fatty acids, triacylglycerols and cholesteryl
esters. The averaged PL/protein ratio of close to one
(Giesing, 1978) is suggestive of a dynamic protein-
lipid interplay serving as the control framework of
specific membrane functions, i.e. neurotransmitter
receptor binding and transmission of signals across
the lipid bilayer. The fluidity of membrane lipids
has been established as an essential determinant in a
variety of membrane-associated events that are common
in all component cells. Chemical effectors of membrane
fluidity are - in order of significance - the level of
cholesterol (Shinitzky and Henkart, 1979), the degree
of saturation of the acyl chains, the level of sphingomyelin (Barenholz and Thompson, 1980), the ratio
phosphatidylethanolamine/phosphatidylcholine (Hirata
and Axelrod, 1980) and the ratio protein/lipid
(Shinitzky and Henkart, 1979). Modulation of these
effectors is achieved within different periods of time
through regulatory signals, e.g. membrane biogenesis
(Giesing and Gerken, 1982, a), stimulatory input
(Giesing and Gerken, 1982, b; Hirata et al., 1981)
and pathological conditions (Bazan, 1970;

[*]Mailing address: A. Nattermann & Cie. GmbH,
                    Nattermannallee 1, D-5000 Köln 30

Lloyd and Beaumont, 1980; Majewska et al., 1981;
Robert et al., 1976; Sun and Leung, 1974). On a time
scale the ambient physical effectors of lipid fluidity,
e.g. temperature, pressure, pH, membrane potential and
Ca^{2+} (Shinitzky and Henkart, 1979) are virtually in-
stantaneous. Yet the analysis of the correlation between
the activity of receptors and lipid fluidity is
hampered by the vagaries associated with the resolution that ensues from assays on the "microscopic"
level by nuclear magnetic resonance, on the "macroscopic"
level by calorimetric and mechanical means and on the
"submacroscopic" level with the aid of spectral probes.
Furthermore, because of the complex mosaic structure
of the lipid bilayer, the asymmetric distribution and
metabolism of PL (van Deenen, 1981) in vitro analysis
of the lipid protein and lipid ligand interplay may lead
to different results as studies in viable cells.
Inasmuch as the physiological relevance of the overt
activity of ligand receptor binding is substantiated
by PL cultures of nervous tissue developing in vitro
the typical heterogenous network of synaptically
coupled nerve cells (Giesing, 1983; Giesing and Gerken,
1982, a, b) represent a valuable tool. It is the
intention of this paper to provide an outline of the
current views on lipids in regard to neurotransmitters,
both excitatory and inhibitory in nature, e.g. amines,
amino acids, peptides, to receptors and to second
messengers. This aim may be helpful to bridge the
gap of contradiction and unawareness between
"receptorologists" and "membranologists".

TOPOLOGY AND METABOLISM OF MEMBRANE PHOSPHOLIPIDS

The transversal distribution of PL classes in
plasma membranes depicts a pattern of considerable
similarity in most of the cells investigated so far
(van Deenen, 1981). PL facing the cytoplasmic side
are phosphatidylserine (PS), phosphatidylethanolamine(PE)
and phosphatidyl-N-monomethylethanolamine (PME). The
external leaflet shows accumulation of sphingomyelin,
phosphatidylcholine (PC) and phosphatidyl-N-dimethyl-
ethanolamine (PDE). The asymmetry of PL is replenished
through a variety of mechanism ensuing from enzymatic
processes. PC, synthesized de novo in the rough endo-
plasmic reticulum, is translocated from the inner to
the outer leaflet through exchange with phosphatidic
acid originating from phospholipase D activity
(DeKruijff and Baken, 1978). Furthermore, the asymmetric
distribution of synthetic pathways encompasses PL

methylation yielding PC (Crews et al., 1980; Hirata et
al., 1979; McGivney et al., 1981). Transbilayer
movement of PL in bilayer vesicles can be induced by
introducing into the matrix different physical
properties between the inner and outer monolayer,
insertion of bilayer spanning proteins and triggering
of the formation of non-bilayer arrangements in the
lipid barrier (van Deenen, 1981). On the whole, the
composition of membraneous PL is a determinant of
fluidity which is an established regulator in membrane
functioning.

MODULATION OF THE CHEMICAL EFFECTORS OF LIPID FLUIDITY

Membrane lipid fluidity is amenable to various
manipulatory procedures that will affect the density or
free volume of lipid domains. In turn, the corresponding lipid microviscosity is altered and physiological
functions which are diffusion-dependent, are modulated
or restored when previously impaired. Invasive
manipulations comprise enzymatic PL-hydrolysis,
extracts with organic solvents or application of
detergents. Non-destructive means are reconstitution
with PL and introduction or exchange of PL.

Invasive techniques have been applied for a
variety of purposes, among them receptor assays, in the
nervous system. Recent results are summarized in
Table 1. Most experiments have been performed in
membrane preparations stemming from donor animals or
post-mortem brain of man. Because one might expect
considerable differences between membranes and
viable cells we have examined GABA receptor binding in
cultures of grey matter of cerebral cortex. The
cultures develop in vitro a complex pattern of bioelec-
tric activity ensuing from highly differentiated
synapses (Giesing, 1978). Comparing the results
between brain and culture membranes it has to be taken
into consideration that the dissection procedure of
brain requires a period of cellular hypoxia and tissue
ischemia.

Polyunsaturated fatty acids will be liberated
from membrane PL in a very rapid process (Bazan, 1970;
Majewska et al., 1981). Because of the free
accessibility to cultured neurons it seems very unlikely
that harvest of the tissue specimens induces
phospholipase activity. Concerning the GABA receptor
a major difference between the sources for applicable

Table 2.   Distribution of dansyl-phosphatidylethanolamine[a]

Cellular Structure	Scanning Fluorimetry	In Vitro Fluorimetry	Subcellular Fraction
	% of total fluorescence		
Nucleoli	5.4	4.5	Nuclei, debris
Surface membrane	19.5	12.0	Crude mito-chondria
Endoplasmic reticulum	43.7	32.3	Microsomes
Cytoplasmic inclusions	31.7	51.2	Cytosol

[a]Scanning fluorimetry was carried out with the aid of a high speed scanning table adapted to a scanning fluorescence light microscope (Zeiss, Oberkochem). In vitro fluorimetry was worked out in lipid extracts of subcellular fractions in chloroform as indicated. Wave length specification: excitation at 352 nm; emission at 517 nm.

Table 3.   Effect of Lipids on Receptor Activity[a]

Lipid	Receptor	Effect	Reference
Ch	nicotinic	preservation	Criado et al.,1982
PC, PS,PE	dopamine sensitive adenylate cyclase	restoration	Anand-Srivastava & Johnson,1981
PE	GABA	inhibition	Toffano et al.,
PS,PC		ineffective	1981
PE,PME	GABA	stimulation	Giesing & Gerken,
PC,PDE		inhibition	1982,a; Giesing, 1983
PS	opiate	enhancement	Abood et al.,1980
		protection	Dunlop et al.,1979
PC		decrease	Heron et al.,1981
CS		enhancement	Law et al.,1978

[a]abbreviations: Ch=cholesterol; CS=cerebroside sulfate; PS=phosphatidylserine; PE=ethanolamine; PME=-N-monomethyl-ethanolamine; PDE=N-dimethylethanolamine; PC=-choline.

specificity of PL action attributing on the whole to
the hydrophobic moiety a masking property for about
75% of the binding capacity, i.e. the sum of operating
and unloaded binding sites. At higher resolution PL in
the cytoplasmic leaflet turned out to stimulate receptor
activity; PL facing the external side inhibit receptor
binding expressed either by a decrease in the values of
B_{MAX} or K_D or both (Giesing, 1983; Giesing and Gerken,
1982, a). Among the latter PL class phosphatidyl-N-
dimethylethanolamine (PDE) plays a dominant role.

PDE formation via the methylation pathway can be
induced through cholinergic stimulation (Fig. 1). The
remaining PL-incorporating CH_3 groups from an exogenous
source exerted reduced activity. The structural
specificity on PDE biosynthesis suggests in brain the
presence of a third methyltransferase catalyzing in a one
step reaction the methylation of phosphatidyl-N-monomethylethanolamine (PME). This is in contrast to
previous findings (Crews et al., 1980). The maximum
stimulation in PDE formation appeared at a range between
10^{-2} and 10^{-3} mol/L of the cholinergic stimulus. It is
interesting to note that cholinergic agonists are considered to be effective at the postsynaptic site in the
millimolar range. The carbachol-induced PL effect was
accompanied by a threefold increase in available high
and low affinity binding sites. The strength of the
binding between ligand and receptor increased steadily
with preceding carbachol incubation.

In more elaborate experiments we have shown that
the receptor per se is not a lipoprotein in nature. PL
govern the activity of one or two modulator proteins
that have inhibitory and stimulatory activity on the
receptor (Giesing, 1983; Giesing and Gerken, 1982, a).
Endogenous modulators have been described elsewhere
(Chiu and Rosenberg, 1981; Guidotti et al., 1978;
Yoneda and Kuriama, 1980). The PL-modulator concept
emerging from our own work can be regarded as a hint for
a molecular explanation of GABA receptor desensitization
and supersensitivity that was previously defined on the
basis of binding activity (Lefkovitz et al., 1977;
Williams and Lefkovitz, 1977; Yoneda and Kuriyama, 1980).
The modulator can mask the GABA binding site (inhibition)
and unmask the binding site (stimulation). The first
step is mediated through PC and PDE or through PE and
PME degradation. Stimulation of ligand binding is
achieved through PE and PME or through hydrolysis
of PC and PDE. The changes in lipid microviscosity
associated with PL biosynthesis and degradation may

account for lateral or vertical displacement on the
modulator and/or the receptor (see next chapter).
The results obtained with cultures of heterogenous
nerve cells show for the first time a functional
interplay of cholinergic and GABA-ergic neurons in
cerebral cortex (Kelly and Moore, 1978; Lewis and
Shute, 1978; Mao et al., 1977; McGeer et al., 1973;
Phillis   1975; Shute and Lewis, 1967; Zsilla et
al., 1976) on a molecular level that may explain also
a variety of pathological conditions (Lloyd and Beaumont,
1980).

PRINCIPLES OF LIPID RECEPTOR INTERACTION

Receptors as other membrane proteins are amenable
to interaction with lipids via electrostatic forces
in the headgroup layer and via hydrophobic bonds in the
hydrocarbon core. Both are determinants of lipid
fluidity. The steady state level of receptors, i.e.
position and distribution, and the dynamic level, i.e.
passive diffusion, ensue from membrane fluidity.
Changes in the receptor-lipid interplay can be measured
through determination of receptor position and activity.
Analysis of protein motility is problematic (Saffmann
and Delbrück, 1975; Galla et al., 1979) since lateral
passive diffusion of proteins in biological membranes
is complied with membrane thickness and protein radius
in addition to lipid viscosity (Anderson and Mago,1980).

Increase in lipid microviscosity decreases the
lipid free volume leading in turn to reduced solubilization of the protein in the hydrocarbon core
(Shinitzky, 1979). As a result the energy of the interaction decreases (Gerson, 1982). The protein will
undergo "vertical displacement" into the aqueous domain
of either membrane side (Borochov and Shinitzky,1976;
Shinitzky, 1979). The alternative and compensatory
mechanism is a shift in the equilibrium position
through "lateral displacement" on the grounds of
protein-protein association. PL-controlled modulator
activity on the GABA receptor, for instance, could
ensue from vertical displacement of the modulator
protein and/or from displacement in the lateral axis
due to receptor-modulator interaction. Lateral displacement can even induce segregation of lipid and
protein domains (Wunderlich et al., 1975; Kleeman and
McConnell, 1976; Cherry et al., 1980).

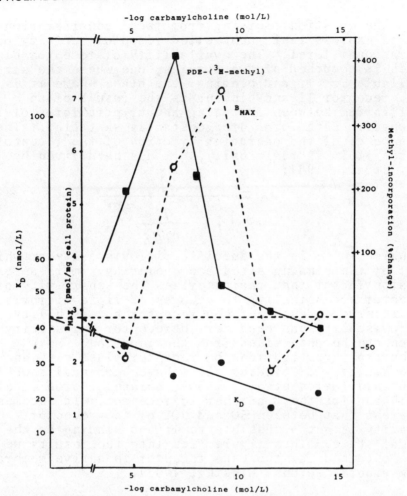

Fig. 1. Effect of cholinergic stimulation on phospholipid methylation and high affinity GABA receptor
binding. The cholinergic stimulus was given to viable
nerve cells for 30 min ranging from 10^{-2} to 10^{-6} mol/L.
Following the stimulus cellular phospholipid methylation
was determined either from exogenous methionine or
S-adenosyl-methionine. The precursors were given at
200 μmol/L for 10 min. The baseline incorporation of
methyl groups into cellular phosphatidyl-N-dimethyl-
ethanolamine (PDE) was 0.93 pmol/mg protein from
methionine and 6.3 pmol from S-adenosyl-
methionine. GABA receptor binding was also measured immediately after the stimulation period. Control values:
B_{MAX} = 2.8 pmol/mg protein; K_D = 43 nmol/L. Averages of
at least 4 determinations are given.

The question emerging from these considerations is
how to express lipid controlled receptor activity on the
operational level. The availability of receptor binding
sites is composed of two domains, one where the site is
available for ligand binding, the other where it is not,
i.e. receptor ligand binding is the result of an
equilibrium between operating and cryptic forms of the
receptor. Both forms are constantly shuttling. The
dependence of the operating fraction of the receptor, α,
on the lipid microviscosity, $\bar{\eta}$, has been given by
Yuli et al., 1981:

$$\alpha = \cfrac{1}{1 + \cfrac{\bar{\eta}}{\bar{\eta}^{1/2}}^{-m}}$$

in which $\bar{\eta}^{1/2}$ is the specific microviscosity at which
half of the binding sites are operating. m is an ex-
pansion factor that characterizes the sensitivity of the
receptor accessibility to changes of lipid microviscosity.
The interdependence between receptor accessibility and
microviscosity has been termed receptor "plasticity"
which can be determined from the equation. More
elaborate investigations on receptor plasticity and
accessibility of binding sites have been published for
serotonin (Heron et al., 1980), opiates (Heron et al.,
1981) and for the ß-adrenergic receptor. It becomes
apparent that between 50 and 90% of the receptor
capacity is not accessible to ligand binding at the
normal state. This figure fits into the results we
have obtained for the GABA receptor in native membranes
from viable neurons (Giesing, 1983).

As in other equilibria, thermal fluctuations around
the median vertical position of the receptor are to be
expected implying that individual receptor molecules spend
varying, microviscosity-dependent periods of time in an
overexposed position. The receptors would then have a
tendency to shift into the aqueous phase where they are
vulnerable to enzymatic degradation or can be shed off
as shown for serotonin (Heron et al., 1980; Shih and
Young, 1978). It is tempting to speculate that failure
of a PL-controlled activity of the GABA receptor account
for diseases such as Huntington's chorea (Lloyd and
Beaumont, 1980) and others. This mechanism might also
be functionally connected with receptor's state of
desensitization and supersensitivity. Finally, pro-
trusion and detachment of receptor molecules from
membranes upon excess rigidification should be a
valuable approach to isolate the proteins.

PRINCIPLES OF LIPID LIGAND INTERACTION

Two facets of lipid ligand interaction become
apparent considering the penetration of the ligand into
the matrix of the membrane and undergoing binding at the
receptor's binding site. The first step involves exposure of the ligand to the lipid moiety to which it may
bind, e.g. serotonin to phosphatidylserine (PS) as
reported by Yandrasitz et al., 1980) and to
gangliosides (Tamir et al., 1980), substance P to
anionic and zwitterionic PL (Lembeck et al., 1979) and
opiates to cerebroside sulfate in a saturable, stereo-
selective fashion (correlating even affinity and
pharmacologicyl potency (Cho et al., 1976, a,b). Opiates,
both agonists and antagonists, increase lipid bilayer
fluidity that in turn can affect the receptor protein
albeit not in parallel to their analgesic effects
(Johnson et al, 1979). The presence of a basic charge
in the ligand molecules seems to be a prerequisite for
binding to lipids. In enkephalin neither the tyrosyl
hydroxyl group nor the C-terminal carboxyl group is
involved in binding to PS (Jarell et al., 1980). The
principal interacting charge is the $-NH_3^+$ group. Upon
binding the internal motion in enkephalin is reduced.
Adoption of a rigid structure in a hydrophobic environment has also been found for ß-endorphin binding to
cerebroside sulfate (Wu et al., 1979). Formation of an
α-helix of this polypeptide in lipid solutions may allow
the NH_2-terminal peptide, i.e. [Met] enkephalin, to bind
to the receptor site more efficiently than in a state of
lesser order.

Binding to the receptor site may follow in some
cases the phase of interacting with lipids. The
available number of receptor sites determines the amount
of activated proteins under saturation conditions.
Transition of the inactive receptors - the binding sites
are unoccupied - to the activated state - the ligand is
bound - is presumably mediated by conformational change of
the protein. The process of transition can in turn
affect the lipid microviscosity in the vicinity of the
protein. The change may even dissipate over the whole
lipid domain. Increase or decrease of lipid microviscosity has been reported for serotonin binding
(Rosenberg, 1979) and occupancy of the nicotinic receptor
(Schneweiss et al.,1979).

Dissipation of the lipid microviscosity altered
through activation of the receptor can initiate a
cascade of events that are associated with lateral or

vertical displacement of other proteins. In a dense
receptor population increasing microviscosity can
aggregate active and inactive receptors. The oligomeric
receptor aggregates have been termed the "unit signal"
of receptor activity (Schlessinger, 1980). Following
signal transmission the microaggregates will either be
internalized as co-aggregates or conveyed by an active
process onto "coated pits" to form large aggregates.
These are internalized, disarmed and eventually di-
gested or re-processed (Goldstein et al., 1979;
Schlessinger, 1980). At present evidence for this
mechanism has been provided for hormone and lipoprotein
receptors. It seems unlikely that neurotransmitter
receptors undergo the same process since the frequent
shuttling between the activated and inactive equilibria
rather requires a less complex mechanism.

On the other hand a considerable number of reports
have confirmed the effects of changes in microviscosity
on the enzyme-mediated production of second messenger
nucleotides following receptor activation. Diffusion-
controlled coupling between the occupied ß-adrenergic
receptor, the GTP binding protein and adenylate cyclase
is a well established example of the transmembraneous
process generating intrinsic activity on the grounds of
signal transduction (Rimon et al., 1978; Tokovsky and
Levitzky, 1978; Hanski et al., 1979). It seems that the
diffusion-controlled collision is abortive as long as the
receptor is not occupied. Coupling between receptor in
the activated form and the enzyme would be effective on
the operational level, i.e. lead to the formation of
cAMP (Anand-Srivastava and Johnson, 1981). Up to now
it is not yet clear whether the rate-limiting collision
is between the receptor and the GTP-binding protein or
between the latter and the adenylate cyclase (Salesse
et al., 1982, b; Tokovsky et al., 1982). Also it seems
contradictory whether increase in membrane fluidization
either through receptor activation or through PL
methylation induced by agonist binding (Hirata et al.,
1979; McGivney et al., 1981; Saavedra, 1980) stimulates
nucleotide formation or impairs coupling of the proteins
(Salesse et al., 1982, a).

CONCLUSION

Synaptic transmission is a process ensuing from the
regulatory potency of the lipid protein membrane. Inasmuch as the postsynaptic side contributes to the regulation of signal transduction, the response is more
complex than to be envisaged on the grounds of the

biomolecular reaction of ligand receptor binding. The
number of activated receptors is much lower than it can
be expected from receptor capacity. The reserve of unoccupied receptors can be shuttled into the active form.
The ratio between the two receptor equilibria is
regulated by PL. In this context the asymmetric
distribution of PL plays the key role. It seems likely
that balancing excitatory and inhibitory inputs requires
a reserve capacity of cryptic forms that undergo
activation upon neuronal stimulus. Hence possible
defects in PL composition and/or metabolism of PL in
either leaflet of the nerve cell membrane may account
for pathological conditions in which the balance of
receptor activity is impaired. Membrane PL govern not
only the density of available binding sites but also the
receptor affinity. This sheds light on further facets
of receptor regulation depicting a lipid theory for
various functional states such as supersensitivity and
desensitization.

REFERENCES

1. L.G. Abood, M. Butler and D. Reynolds, Effect of
    calcium and physical state of neutral membranes on
    phosphatidylserine requirement for opiate binding,
    Mol. Pharmac. 17:290 (1980).
2. M.B. Anand-Srivastava and R.A. Johnson, Role of
    phospholipids in coupling of adenosine and dopamine
    receptors to striatal adenylate cyclase, J.Neurochem.
    36, 5:1819 (1981).
3. G.R. Anderson and R.M. Mazo, Models for boundary
    effects on molecular rotation in membranes,
    Biopolymers 19, 1597 (1980).
4. T.J. Andreasen, D.R. Doerge and M.G. McNamee,
    Effects of phospholipase A2 on the binding and ion
    permeability control properties of the acetylcholine
    receptor. Arch.Biochem.Biophys. 194,2:468 (1979).
5. Y. Barenholz and T.E. Thompson, Sphingomyelin in
    bilayers and biological membranes, Biochim.
    Biophys. Acta 604:129 (1980).
6. N.G. Bazan, Effects of ischemia and electro-
    convulsive shock on free fatty acid pool in the
    brain, Biochim. Biophys. Acta 218:1 (1970).
7. J.M. Bidlack and L.G. Abood, Solubilization of
    the opiate receptor, Life Sci. 27:331 (1980).
8. H. Borochov and M. Shinitzky, Vertical displacement
    of membrane proteins mediated by changes in micro-
    viscosity, Proc. Natl. Acad. Sci. USA 73:4526 (1976).

9. R.J. Cherry, U. Müller, C. Holenstein and M.P. Heyn,
   Lateral segregation of proteins induced by
   cholesterol in bacteriorhodopsin-phospholipid
   vesicles, Biochim. Biophys. Acta 596:145 (1980).
10. T.H. Chiu and H.C. Rosenberg, Endogenous modulator
    of benzodiazepine binding in rat cortex, J.
    Neurochem. 36:336 (1981).
11. T.M. Cho, J.S. Cho and H.H. Loh, ^{3}H-cerebroside
    sulfate redistribution induced by cation, opiate
    or phosphatidylserine, Life Sci. 19:117 (1976, a).
12. T.M. Cho, J.S. Cho and H.H. Loh, A model system
    for opiate-receptor interactions: mechanisms of
    opiate-cerebroside sulfate interaction, Life Sci.18:
    231 (1976, b).
13. F.T. Crews, F. Hirata and J. Axelrod, Phospholipid
    methyltransferase in synaptosomal membranes ,
    Neurochem. Res. 5,9:983 (1980).
14. M. Criado, H. Eibl and F.J. Barrantes, Effect of
    lipids on acetylcholine receptor. Essential need
    of cholesterol for maintenance of agonist-induced
    state transitions in lipid vesicles. Biochemistry
    21:3622 (1982).
15. L.M. van Deenen, Topology and dynamics of phos-
    pholipids in membranes. FEBS Letters 123, 1:3(1981).
16. C.E. Dunlap III, F.M. Leslie, M. Rado and B.M. Cox,
    Ascorbate destruction of opiate stereospecific
    binding in guinea pig brain homogenate, Mol.Pharmac.
    16:105 (1979).
17. S.J. Enna and S.H. Snyder, Influences of ions,
    enzymes and detergents on y-aminobutyric acid
    receptor binding in synaptic membranes of rat brain,
    Molec. Pharmac. 13:442 (1977).
18. H.J. Galla, W. Hartmann, U. Theilen and E. Sackmann,
    On two-dimensional passive random walk in lipid
    bilayers and fluid pathways in biomembranes,
    J. Membrane Biol. 48, 215 (1975).
19. D.F. Gerson, Interfacial free energies of the
    positioning and aggregation of membrane proteins,
    Biophys. J. 37: 145 (1982).
20. C.T. Giambalvo and P. Rosenberg, The effect of
    phospholipases and proteases in the binding of
    y-aminobutyric acid to junctional complexes of
    rat cerebellum, Biochim. Biophys. Acta 436:741
    (1976).
21. M. Giesing, Explantatkulturen des Nervensystems:
    Ein neues Modell für die Neurochemie. Bericht von
    der Regulation einiger Lipidbausteine, Habilitations-
    schrift, Universität Bonn, 1978.

22. M. Giesing, Organized cultures of nerve tissue. A novel model system for studies of lipid protein interaction on the functional level, in: "Physical Methods in Biological Membranes and their Model Systems: Possibilities and Limits", F. Conti, W. Blumberg, J. de Gier, F. Pocchiari eds. (1983), Plenum Press (in press).

23. M. Giesing and U. Gerken, The effects of carbamyl-choline on extrasynaptic phosphatidylcholine bio-synthesis in grey matter of cerebral cortex, in: "Phospholipids in the Nervous System", G.B. Ansell, L.A.Horrocks and G. Porcellati eds., Raven Press, New York (in press) (1982, a).

24. M. Giesing and U. Gerken, The role of asymmetrically distributed phospholipids in the binding of gamma-aminobutyric acid, in: "Basic and Clinical Aspects of Molecular Neurobiology", A.M. Giuffrida Stella, G. Gambos, G. Benzi and H.S. Bachelard, eds., Fonda-zione Internazionale Menarini, Milano, Italy,p.135 ff. (1982, b).

25. M. Giesing and F. Zilliken, Analysis of lipid compo-nents in organotypic cultures of cerebellum during development, Brain Research 111:212 (1976).

26. J.L. Goldstein, R.G.W. Anderson and M.S. Brown, Coated pits, coated vesicles and receptor-mediated endocytosis, Nature 279:679 (1979).

27. A. Guidotti, G. Toffano and E. Costa, An endogenous protein modulates the affinity of GABA and benzodia-zepine receptors in rat brain, Nature 275:553 (1978).

28. F. Hanski, G. Rimon and A. Levitzki, Adenylate cyclase activation by the ß-adrenergic receptors as a diffusion-controlled process, Biochemistry 18:846 (1979).

29. D.S. Heron, M. Shinitzky, M. Hershkovitz and D. Samuel, Lipid fluidity markedly modulates the binding of serotonin to mouse brain membranes, Proc.Natl. Acad.Sci. USA 77:7463 (1980).

30. D.S. Heron, M. Israeli, M. Hershkovitz, D. Samuel and M. Shinitzky, Lipid-induced modulation of opiate receptors in mouse brain membranes, Eur.J. Pharmacol. 72:361 (1981).

31. F. Hirata and J. Axelrod, Phospholipid methylation and biological signal transmission, Science 209:1082 (1980).

32. F. Hirata, W.J. Strittmatter and J. Axelrod, ß-adren-ergic receptor agonists increase phospholipid methyl-ation, membrane fluidity and ß-adrenergic receptor-adenylate cyclase coupling, Proc.Natl.Acad.Sci.USA 76:368 (1979).

33. H.C. Jarell, R. Deslauriers, W.H. McGregor and
    J.C.P. Smith, Interaction of opiate peptides with
    model membranes. A carbon-13 nuclear magnetic study
    of enkephalin binding to phosphatidylserine,
    Biochemistry 19, 385 (1980).
34. D.A. Johnson, R. Cooke, H.H. Loh, Effect of opiate
    agonists and antagonists on lipid bilayer fluidity,
    Mol. Pharmacol. 16: 154 (1979).
35. P.H. Kelly and K.E. Moore, Decrease of neocortical
    choline acetyltransferase after lesions of the
    globus pallidus in rats, Exp. Neurol.61:475 (1978).
36. W. Kleemann and H.M. Cornell, Interactions of
    proteins and cholesterol with lipid bilayer membranes,
    Biochim. Biophys. Acta 419:206 (1976).
37. B. De Kruijff and P. Baken, Rapid transbilayer
    movement of phospholipids induced by an asymmetrical
    perturbation of the bilayer, Biochim. Biophys.Acta
    507:38 (1978).
38. P.Y. Law, R.A. Harris, H.H. Loh and E.L. Way,
    Evidence for the involvement of cerebroside sulfate
    in opiate receptor binding: studies with azure a
    and jimpy mutant mice, J. Pharmacol. Exp. Therap.
    207:458 (1978).
39. R.J. Lefkovitz, D. Mullikin, C. Wood, T. Goore and
    C. Mukherjee, Regulation of prostaglandin receptors
    by prostaglandins and guanine nucleotides in frog
    erythrocytes, J. Biol. Chem. 252:5295 (1977).
40. F. Lembeck, A. Saria and N. Mayer, Substance P:
    model studies of its binding to phospholipids,
    Naunyn-Schmiedeberg's Arch. Pharmacol. 306:189 (1979).
41. P.R. Lewis and C.C.D. Shute, Cholinergic pathways
    in the CNS in:"Handbook of Psychopharmacology, Vol.9:
    Chemical Pathways in the Brain", L.L. Iversen,
    S.D. Iversen and S.H. Snyder eds., p. 315 ff. (1978),
    Plenum Press, New York.
42. K.G. Lloyd and K. Beaumont, Possible role of
    phospholipids in GABA receptor function in human and
    rat brain, Brain Res. Bull. 5:285 (1980).
43. K.G. Lloyd and L. Davidson, ^{3}H-GABA binding in brains
    from Huntington's chora patients: altered regulation
    by phospholipids? Science 205:1147 (1979).
44. E.G. McGeer, H.C. Fibiger, P.L. McGeer and S.Brooke,
    Temporal changes in amine synthesizing enzymes of
    rat extrapyramidal structures after hemitransections
    or 6-hydroxydopamine administration, Brain Res.
    52:289 (1973).

45. A. McGivney, F.T. Crews, F. Hirata, J. Axelrod and R.P. Siragavian, Rat basophilic leukemia cell lines defective in phospholipid methyltransferase enzymes, Ca^{2+} influx and histamine release: reconstitution by hybridization, Proc. Natl.Acad.Sci. USA 78:6176 (1981).

46. M.D. Majewska, R. Manning and G.Y. Sun, Effects of postdecapitative ischemia on arachidonate release from brain synaptosomes, Neurochem. Res. 6:567 (1981).

47. C.C. Mao, G. Marco, A. Revuelta, L. Bertilsson and E. Costa, The turnover rate of y-aminobutyric acid in the nuclei of telencephalon: implications in the pharmacology of anti-psychotics and of a minor tranquilizer, Biol. Psychiat. 12:359 (1977).

48. J.W. Phillis, Evidence for cholinergic transmission in the cerebral cortex, in:"Neurohumoral Coding of Brain Function", R.D. Myers and R.R. Drucker-Colin, eds., Plenum Press, New York, p.57 ff. (1975).

49. G. Rimon, E. Hanski, S. Braun und A. Levitzki, Mode of coupling between hormone receptors and adenylate cyclase elucidated by modulation of membrane fluidity, Nature 276:394 (1978).

50. P.H. Rosenberg, Effects of halothane, lidocaine and 5-hydroxytryptamine on fluidity of synaptic plasma membranes, myelin membranes and synaptic mitochondrial membranes, Arch. Pharmacol. 307:199 (1979).

51. J.M. Saavedra, Increased adrenaline, ß-adrenoreceptor stimulation and phospholipid methylation in pineal gland of spontaneously hypersensitive rats, Clinical Science 59:239 (1980).

52. P.G. Saffmann and M. Delbruck, Brownian movement in biological membranes, Proc. Natl. Acad. Sci. USA 72:3111 (1975).

53. R. Salesse, J. Garnier and D. Daveloose, Modulation of adenylate cyclase activity by the physical state of pigeon erythrocyte membrane. 2. Fluidity-controlled coupling between the subunits of the adenylate cyclase system, Biochemistry 21:1587 (1982, b.).

54. R. Salésse, J. Garnier, F. Leferrier, D. Daveloose and J. Viret, Modulation of adenylate cyclase activity by the physical state of pigeon erythrocyte membrane. 1. Parallel-drug-induced changes in the bilayer fluidity and adenylate cyclase activity, Biochemistry 21:1581 (1982, a.).

55. J. Schlessinger, The mechanism and role of hormone-induced clustering of membrane receptors, Trends Biochem. Sci. 5:210 (1980).

56. F. Schneweiss, D. Naquira, K. Rosenbeck and A.S. Schneider, Cholinergic stimulants and excess potassium ion increase in the fluidity of plasma membranes isolated from adrenal chromaffin cells, Biochim. Biophys. Acta 555:460 (1979).

57. G.Y. Sciu and B.S. Leung, Phospholipids and acyl groups of subcellular fractions from human intra-cranial tissues, J. Lipid Res. 15:423 (1974).

58. J.C. Shi and H. Young, The alteration of serotonin binding sites in aged human brain, Life Sci. 23:1441 (1978).

59. M. Shinitzky, The concept of passive modulation of membrane responses, Dev. Cell Biol. 4:173(1979).

60. M. Shinitzky and P. Henkart, Fluidity of cell membranes: current concepts and trends, Intl. Rev. Cytol. 60:121 (1979).

61. C.C.D. Shute and P.R. Lewis, The ascending cholinergic reticular system: neocortical, olfactory and sub-cortical projections, Brain 90:497 (1967).

62. H. Tamir, W. Brunner, D. Casper and M.M. Rapport, Enhancement by gangliosides of the binding of serotonin to serotonin binding protein, J. Neurochem. 34:1719 (1980).

63. G. Toffano, C. Aldinio, M. Bolzano, A. Leon and G. Savoini, Regulation of GABA receptor binding to synaptic plasma membrane of rat cerebral cortex: the role of endogenous phospholipids, Brain Res. 222:95 (1981).

64. A.M. Tolkovsky, S. Braun and C.A. Levitzki, Kinetics of interaction between ß-receptors, GPT protein and the catalytic unit of turkey erythrocyte adenylate cyclase, Proc. Natl. Acad. Sci. USA 79:213 (1982).

65. A.M. Tolkovsky and A. Levitzki, Mode of coupling between the ß-adrenergic receptor and adenylate cyclase in turkey erythrocytes, Biochemistry 17:3795 (1978).

66. L.T. Williams and R.J. Lefkovitz, Slowly re-versible binding of catecholamine to a nucleotide-sensitive state of the ß-adrenergic receptor, J. Biol. Chem. 252:7207 (1977).

67. D.R. Wong and J.S. Horng, Na^+-independent binding of GABA to the Triton-X-100 treated synaptic membranes from cerebellum of rat brain, Life Sci. 20:445 (1977).

68. C.S.C. Wu, N.M. Lee, H.H. Loh, J.T. Yang and C.H. Li, ß-endorphin: formation of α-helix in lipid solutions, Proc. Natl. Acad.Sci. USA 76:3656 (1979).
69. F. Wunderlich, V. Ronai, J. Speth, J. Seelig and A. Blumen, Thermotropic lipid-clustering in tetrahymana membranes, Biochemistry 14:3730 (1975).
70. J.R. Yandrasitz, R.M. Conin, B. Masley and D. Rowe, Evaluation of the binding of serotonin by isolated CNS acidic lipids, Neurochem. Res. 5:465 (1980).
71. Y. Yoneda and K. Kuriyama, Presence of a low molecular weight endogenous inhibitor on ^{3}H-muscimol binding in synaptic membranes, Nature 285:670 (1980).
72. J. Yuli, W. Wilbrandt and M. Shinitzky, Glucose transport through cell membranes of modified lipid fluidity, Biochemistry 70:4250 (1981).
73. G. Zsilla, D.L. Cheney, G. Racagni and E. Costa, Correlation between analgesia and the decrease of acetylcholine turnover rate in cortex and hippo-campus elicited by morphine, meperidine, Vivimol^R and azidomorphin, J. Pharmac. Exp. Ther. 199:662 (1976).

INCORPORATION OF RECEPTORS INTO PLANAR LIPID BILAYERS:

IONIC TRANSPORT PROPERTIES

Franco Gambale,[a] Cesare Usai,[a]

Mauro Robello,[b] and Carla Marchetti[a,c]

a) Istituto di Cibernetica e Biofisica, Corso
   Mazzini 20, 16032 Camogli, Genova (Italy)
b) Istituto di Scienze Fisiche-Università di
   Genova, Via Dodecaneso 33, 16146 Genova (Italy)
c) FIDIA Research Laboratories, Via Ponte della
   fabbrica 3a, 10116 Abano Terme, Padova (Italy)

INTRODUCTION

Plasma membrane separating the cytoplasm from the surrounding medium plays a fundamental role in maintaining the homeostatic equilibrium of the cell. In fact, it is now generally recognized that cell membrane dynamically interacts with external components such as ions, molecules, and other cell membranes.

In 1962 Mueller and his coworkers were able to obtain planar artificial lipid bilayers whose properties were very similar to those of native membranes. Since then the interest of many researchers in black lipid membranes (BLM) has increased more and more. The thickness, electrical capacitance, water permeability, index of refraction, and dielectric breakdown of the model membranes are of the same order of magnitude as those of biological membranes. On the other hand, compared to cell membranes, bilayers made of pure lipids are nearly nonconducting systems. It has been demonstrated, however, that several compounds such as cyclic antibiotics, polypeptides, and proteins are able to increase the ionic permeability of artificial bilayers by many orders of magnitude. In the absence of such intermediates, inorganic ions do not penetrate the lipid bilayers. The same behavior characterizes the transport of more complex molecules unless they present a highly hydrophobic exterior.

On the other hand, planar bilayers present some advantages in view of the complexity of biological membranes. Chemicophysical

parameters characterizing the system (e.g., external voltage, temperature, ionic concentration, pH, composition of the film) can be
controlled very easily. Experiments can be performed in either sym-
metric or asymmetric conditions. Moreover, compounds interacting
with the bilayer can be added in the proper concentration at both
sides of the membrane.

The incorporation into planar bilayers of membrane receptors
and the measurement of ionic transport properties may answer some
questions concerning the transduction of signals following the
interaction with specific ligands. In fact, a modification of the
ionic permeability, which is obvious for receptors associated with
neurotransmitters, seems to characterize also the interaction between hormone receptors and their ligands. Calcium certainly plays
a significant role as secondary messenger (Rasmussen, 1975; Gombert,
1976). Other ions, such as sodium and potassium, influence some
metabolic pathways and contribute to the homeostatic equilibrium of
the cell.

PROCEDURES FOR THE FORMATION OF PLANAR BILAYERS

In the last twenty years several authors have developed different experimental procedures to form planar bilayers. The first
successful attempt - still used today with minor modifications -
should be ascribed to Mueller et al. (1962). A different method was
developed in 1972 by Montal and Mueller and recently modified by
Schindler (1978). A further modification of the Montal technique
was proposed in 1982 by Vodyanoy and Murphy.

Mueller Technique, Conductance and Capacitance Measurements

In the Mueller technique lipid bilayers are formed onto a hole
(diameter 0.4-2 mm) dividing two compartments of a teflon cell
(Fettiplace, 1975; Jones, 1975; Ti Tien, 1974). In order to stabilize the membranes, the hole border is usually pretreated by
spreading a small amount of lipid dispersed into a hydrocarbon solvent. Then the water solution is symmetrically added to the two
sides of the cell, and the membrane is formed by introducing a few
microliters of the lipid-solvent mixture into the hole through a
Pasteur pipette. n-Alkane molecules with chain length greater than
10 carbon atoms are generally used because it has been shown that
the percentage of solvent molecules present in the lipid bilayer
will be lower the longer the chain of the hydrocarbon (Fettiplace,
1975). In fact, lipids dissolved in squalene - a liquid long-chain
($C_{30}H_{50}$) alkyl molecule - form membranes almost free of solvent
molecules (White, 1978). It has been demonstrated that the presence
of hydrocarbon molecules has no significant influence on the ionic
transport properties of BLM.

The initial macroscopic droplet thins out to form a double
layer, with the hydrophilic heads in contact with the external
aqueous solutions and the hydrophobic tails oriented toward the
interior of the film. Both Van der Waals forces and polar interactions help to stabilize the membrane (Gabler, 1978), which is
anchored to the teflon support by a very thick torus, the Plateau-
Gibbs border.

The formation of the lipid bilayer can be followed by illumina-
ting the system with a white light source. Colored interference
fringes are present until the lipid droplet remains thick. Black
spots, rapidly invading the entire membrane surface, reveal the
formation of a lipid double layer.

Particular care should be taken to reduce vibrational and electrical shocks. In fact mechanical vibrations may cause either membrane instability or microphonic noise which can mask the actual
membrane signal. Moreover, owing to the high resistance of "non-
doped" BLM, electrical noise can be very large. To reduce vibrational and electrical disturbances, the cell is enclosed in a
Faraday cage and the entire system placed on an antivibrational
support.

Conductance is measured by applying a step voltage and recording
the ionic current flowing through the membrane. Two Ag-AgCl elec-
trodes are generally used, and experiments are usually performed
keeping the potential applied to the bilayer constant (voltage clamp
technique). The simplest circuit to clamp a lipid bilayer at a constant potential can be obtained by placing a small resistance in
series with the membrane; the current and the applied voltage can be
measured with a voltmeter or an operational amplifier. More complex
circuits, including a high-impedance operational amplifier in a
zero-ground configuration, should be used to measure sharp and small
signals (Fettiplace, 1975).

The equivalent circuit for the cell (Fig. 1) is represented by
the membrane resistance R_m in parallel with the membrane capacitance
C_m in series with the aqueous solution and electrode resistance r_s
(Feldberg, 1978; Fettiplace, 1975).

As already mentioned, the resistance of bilayers made from pure
lipids is very high, of the order of 10^8-10^{10} Ω cm^2, and it can be
lowered by many orders of magnitude by adding to the system different compounds which are able to transport ions.

Capacitance measurements give information on the membrane
thickness. In fact, on the assumption that the lipid bilayer behaves as a parallel-plate capacitor, the capacitance of a membrane
of surface A will be given by

$$C_m = \epsilon_0 \epsilon_r A/d$$

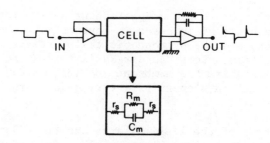

Fig. 1.   Schematic circuit for the measurement of electrical pro-
          perties of planar bilayers.

where $\varepsilon_0 = 8.85 \cdot 10^{-14}$ F/cm is the dielectric constant in vacuum,
$\varepsilon_r \simeq 2$ the relative dielectric coefficient, and d the thickness of
the membrane hydrocarbon core. A specific capacitance of the order
of 0.8 μF/cm^2 was measured for solvent-depleted lipid bilayers,
from which a value of d of the order of $\simeq$ 25 Å can be derived.
Bilayers formed from lipids dissolved in hydrocarbon solvents of
short chain length will be characterized by a higher value of C_m
(0.45 μF/cm^2), thus indicating a greater thickness of the membrane
due to the presence of solvent molecules in the membrane phase.

Montal and Related Techniques

The formation of planar lipid bilayers by apposition of two
monolayers (originally described by Takagy et al., 1965; Ti Tien,
1975) was developed in 1972 by Montal and Mueller. The system consists of two monolayers, at an air-water interface, in a teflon
cell divided by a thin septum (25 μm) containing a small hole ($\simeq$
200 μm) above the water level. After the two monolayers are spread
on both sides of the septum, the water level is raised first on one
side and then on the other. A first monolayer is formed through
the hole when one water level is raised (Fig. 2); then the lipid
bilayer is formed by apposition of the hydrocarbon chains of the
second monolayer by raising the water level in the second chamber.

These bilayers are almost free of solvent molecules because highly
volatile solvents are used to disperse the monolayers over the water
solution. The formation of the membrane is controlled by monitoring
capacity current in response to a voltage step. The electrical resistance is in the range 10^8–10^{10} Ω cm^2, while the specific capacitance is 0.8 μF/cm^2. Planar bilayers obtained by using the Montal
technique are functionally identical to bilayers formed by the
Mueller method with squalene or n-hexadecane as solvent (Montal,

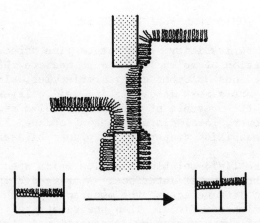

Fig. 2. Formation of a lipid bilayer by the apposition of two
monolayers (adapted from Montal and Mueller, 1972).

1972; White, 1978), but in this case asymmetric bilayers can be
formed from monolayers of different composition.

A further improvement of the Montal technique was achieved by
spreading on the water-air interface single vesicles which may contain proteins from different sources (Montal et al., 1977; Schindler
et al., 1978). Proteins present in the monolayers will be transferred into the bilayer by apposition of the hydrophobic regions of
the two monolayers, as already described. Moreover formation of
bilayer leaflets may be obtained by natural proteolipid vesicles
which were never treated with detergents. Organic solvents are not
used in this method, which seems to add the advantages of planar
bilayer techniques to the completeness of natural membranes.

The formation of solvent-free planar bilayers by a further
modification of the Montal-Mueller technique was recently described
by Vodyanoy and Murphy (1982). This method enables one to form very
stable leaflets of large surface area. A hole (up to 2 mm) is
drilled into the wall of a very small teflon chamber (10 µl in volume). The microscopic cell filled with water solution is raised
from below and dipped again through a lipid monolayer spread on the
water-air interface of a large chamber which constitutes the second
compartment of the system. An interesting characteristic of this
method seems to consist in the higher mechanical stability of the
bilayer, which must be ascribed to the fact that small modifications
of the pressure in the external compartment will be equilibrated by
the pressure in the closed microscopic compartment.

Values of the electrical resistance and capacitance are very
close to values for solvent-free lipid bilayers obtained by other
methods.

MODIFICATION OF IONIC TRANSPORT PROPERTIES

Electrical permeability measurements give direct information on the incorporation of molecules able to mediate the transport of either inorganic ions or other charged molecules. Ions or other compounds which are almost insoluble in lipid bilayers can cross the membrane in the presence of certain molecules as carriers or channels. Moreover, local disturbances of ordered lipid matrix may modify some permeability properties of planar bilayers.

Carriers are lipid-soluble molecules which specifically bind ions at the membrane-water interface. A charged complex is formed which is able to diffuse across the membrane upon the application of an external electric field. Then the substrate is released at the other side of the membrane by dissociation of the complex. It has been shown that, among other compounds, the macrotetralide actins and valinomicin act as carriers forming a specific complex (1:1) with some cations. Owing to the larger radius of the charged complex and the external hydrophobic coat, a decrease in the membrane energy barrier is induced (McLaughlin, 1977; Szabo et al., 1973).

The formation of a conducting pore may be consequence of lateral aggregation of molecules which cooperate to form a hydrophilic pathway spanning the membrane. This seems to be the case when alamethicin molecules are inserted into artificial lipid bilayers (Hladky and Haydon, 1972). An ionic channel may be present in the structure of a single molecule or, as happens for Gramicidin A, two monomers associate head to head to form a helical dimeric pore (Finkelstein et al., 1981; Urry, 1971; Bamberg et al., 1976). The Gramicidin monomer is formed by fifteen alternate L and D hydrophobic amino acids whose apolar side chains face the exterior of the channel, while the hydrophilic interior, almost ideally permeable to cations, is formed essentially by the oxygens of the carbonyl groups (Finkelstein, 1981).

It has been shown that other antibiotics like monoazomycin, nystatin, and amphotericin form ionic pores. Moreover, other molecules such as EIM - excitability inducing material of bacterial origin - (Bean, 1969), hemocyanin obtained from various molluscans (Alvarez, 1975; Menestrina and Antolini, 1982), black widow spider venom (Finkelstein, 1976), and other proteinaceous compounds are able to form ionic channels when incorporated into planar bilayers.

The amplitude of the channel current and the duration time depend on the lipid used to form the bilayer, on the ionic concentration, and, in some cases, on the applied voltage. For example, the unit conductance of a Gramicidin A channel (Fig. 3) is of the order of 20 pS (1 pS = 10^{-12} Ω^{-1}), while the conductance of a single EIM pore was found to be of the order of 400 pS.

The current increase induced by the opening of a single channel is of the order of 10^{-12} A if the pore conductance is 10 pS and
if a 100 mV voltage is applied to the membrane. This means that
about $6 \cdot 10^6$ ions/sec will flow through the channel. On the other
hand, it is well known that the insertion of a carrier into the
bilayer will move about 10^3 ions/sec. This means that each anti-
biotic molecule is able to induce current fluctuations of about
10^{-16} A (Blumental, 1979). Therefore, at high bilayer resistance,
present electronic devices can resolve single fluctuations induced
by small channels ($\simeq$ 1 pS), while the current increase induced by
the insertion of one carrier molecule cannot be resolved. As a
consequence, compounds acting as channels will show typical step-
like current records, while carriers determine an apparent smooth
conductance increase.

In some cases insertion of molecules - or molecular complexes -
causes local disruption of the ordered lipid structure and therefore
induces a transient and noisy increase in the ionic current. The
local disturbance may also give rise to channel-like conductance
changes. These channels can be easily distinguished from "real"
pores because they do not present steps of uniform size. Moreover,
the incorporation of compounds disordering the liquid-crystalline
structure generally leads to instability of the membrane (Blumental,
1979).
molecules previously inserted into the lipid film may give rise to
Also, interactions between external ligands and receptor molecules previously inserted into the lipid film may give rise to
carrier or channel transport mechanisms or induce local disruption
of the membrane structure.

MODIFICATIONS OF IONIC PERMEABILITY INDUCED BY RECEPTOR
INCORPORATION: INTERACTION WITH EXTERNAL LIGANDS

Since 1970 many papers have been published on the incorporation of proteins and membrane receptors into planar bilayers; induced ionic permeability modifications have been extensively analyzed (for complete reviews, see: Blumental, 1979; Montal et al.,
1981).

Since our aim is to give some basic ideas on the usefulness of
the BLM technique in studying interactions between receptors and

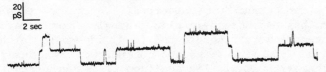

Fig. 3.   Current fluctuations of a dioleoylphosphatidylcholine
          membrane in the presence of very small amounts of Grami-
          cidin, 1 M KCl. Applied potential 60 mV, T = 25°C.

their ligands, we will limit ourselves to relatively few examples
concerning the insertion of acetylcholine receptors and the incorporation of membrane glycolipids and their interactions with toxins,
hormones, and lectins. Translocation of proteins following the
binding with specific receptors involved in endocytosis will also
be discussed.

Acetylcholine Receptors

The binding of acetylcholine (AcCho) to postsynaptic acetylcholine receptors (AcChoR) induces the opening of cation-selective
channels which transduce a chemical signal into an electric one
(Katz, 1966). Neher and Sackmann gave the first direct evidence
that AcCho is able to induce square pulse-like current events at
the neuromuscular junction of frog skeletal muscle fiber (Neher
and Sackmann, 1976).

The reconstitution of AcChoR functions has been studied either
in lipid vesicles (Cartaud et al., 1980; Lindstrom et al., 1980;
Wu et al., 1981) or in planar bilayers (Parisi et al., 1971; Kemp
et al., 1973; Reader et al., 1974; Goodal et al., 1974; Schindler
and Quast, 1980; Nelson et al., 1980; Boheim et al., 1981).

Goodal et al. (1974) were able to show the opening of single
channels incorporating into BLM an impure acetylcholine receptor
from mouse brain in the presence of the AcCho antagonist carbachol.
The channel was permeable to either sodium or chloride ions. Two
different conductance values were measured for both ions: 59 and
240 pS for Cl^-, 37 and 150 pS for Na^+. Tubocurare and atropine
lowered the amplitude and the occurrence of single channels.

Because of improvements in the purification procedure of AcChoR
and in the techniques of reconstituting in planar bilayers the functional properties of native membranes, most of the successful experiments have been performed in the last two years. Schindler and
Quast (1980) formed planar bilayers from vesicles of the Torpedo
marmorata electric organ without detergent treatment. Acetylcholine
receptors remained in their native environment and the bilayer was
formed by apposition of two monolayers obtained by different ratios
of vesicles containing AcChoR to pure phospholipid vesicles. Membranes containing different amounts of receptor molecules can thus
be formed. In the absence of specific ligands no modifications of
the membrane conductance were observed. In the presence carbamoylcholine (CbmCho) and succinylcholine, single channels or macroscopic
current increases were recorded and found dependent on the receptor
concentration. The ligand-induced conductance is saturated above
1 µM of carbamoylcholine. The addition of d-tubocurarine inhibited
the effects of carbamoylcholine and succinylcholine. Pre-incubation
in the presence of α-bungarotoxin suppressed any conductance increase induced by the agonists.

Single channels were characterized by a mean open time of 1.3 ms and a conductance of 90 ± 10 pS, at 100 mV and 1 M NaCl. A slight cationic selectivity was verified:

$$PNa^+/PCl^- = 8 \pm 0.4$$
$$PK^+/PCl^- = 6.8 \pm 0.5$$

Very similar results were obtained by Nelson et al. (1980), who reconstituted in phospholipid bilayers functional channels associated with purified AcChoR from Torpedo californica electric organ. The authors purified the receptor by cholate dialysis in the presence of soybean lipids. The preparation contained only the four subunits $\alpha,\beta,\gamma,\delta$. In this case too the membrane was formed by apposition of two monolayers obtained by spreading soybean lipid vesicles containing AcChoR and supplementary lipid vesicles free of receptors at the air-water interface. Different conductance increases were induced by addition of different concentrations of CbmCho. A large increase in the membrane conductance was followed by a rapid desensitization; this process can be enhanced by additional administration of CbmCho at the peak response. This is in accordance with a conformational change of the receptor, following the binding of the agonist. Asymmetric membranes, presenting AcChoR at one side, develop asymmetric responses, while small amounts of curare (1.25 µM) inhibit the response of carbamoylcholine (0.2 µM). By reducing the number of receptors present in the membrane phase the opening of single channels was observed. The mean conductance was estimated to be 16 ± 3 pS in NaCl = 100 mM, applied potential V = 10 mV, duration 35 ± 5 msec.

More recently, Boheim et al. (1981) were able to confirm that the ionic channel is formed by one or more subunits participating in the receptor monomer. In fact no differences were observed in the channel pattern developed by AcChoR-rich membrane fragments and those depleted of non-receptor peripheral peptides. The authors were able to fuse lipid vesicles containing receptors from Torpedo marmorata electric organ with preformed (Montal-Mueller) lipid bilayers made from 1-stearoyl-3-miristoylglycero-2-phosphocholine. In the presence of different agonists (AcCho, CbmCho, and suberyl-dicholine) and below the phase transition temperature of the lipid (26°C) they recorded opening of single channels very similar to those observed in native systems. The usual concentration of AcCho was 0.4 µM, but effects were also observed at 10 nM. The conductance of the single channel (mean value ≈ 95 pS) did not depend on the applied potential, as shown by the linear current-voltage relationship obtained with 1 M NaCl. Also in this case α-bungarotoxin completely abolished the effects induced by AcCho. Ca^{2+} and Mg^{2+} did not modify the conductance of AcCho-induced channels but were able to decrease the conductance of channels which open on addition of CbmCho.

Glycolipids as Membrane Receptors

Glycoproteins and glycolipids are ubiquitous components of cell membranes. Most of them are present in the outer leaflet of of plasma membranes, while the cytoplasmic side is devoid of carbohydrates. Therefore glycoconjugates are claimed to play a relevant role as surface receptors in cell-cell interactions and recognition.

Subunits participating in the AcChoR monomer, glycophorin and band-3 glycoprotein of erythrocytes, insulin receptors of liver and fat cells, receptors for asialoglycoproteins and adrenocorticotropic hormones are glycoproteins which also interact with lectins of different origins.

Also, glycosphingolipids have been shown to interact with different biofactors such as bacterial toxins, neurotoxins, glyco-protein hormones, and lectins. The human blood group specificity ABO depends on the carbohydrate moiety linked to glycolipids (Hughes, 1975).

Different classes of glycosphingolipids with basically different carbohydrate core structure have been identified. They are based on the sphingosine structure and, while one hydrophobic chain is predominantly a 15:1 hydrocarbon, the second fatty acid chain, which is variable in length and degree of saturation, is linked to the glycerol by an amide group (Ledeen and Yu, 1973).

Gangliosides are complex glycolipids which may present up to nine carbohydrate units. Their oligosaccharide chain is characterized by N-acetyl-neuroaminic acid (or sialic acid), which is responsible for a large part of the electronegative charge at the surface of different cells. It has been shown that the monosialo-ganglioside GM_1 is able to interact with tetanus toxin and cholera toxin (Hakomori, 1980; O'Keefe and Cuatrecasas, 1978). Gangliosides are able to inhibit the binding of such glycoprotein hormones as thyrotropin, chorionic gonadotropin, and luteinizing hormone (Hakomori, 1980).

Despite their relative simplicity, glycolipids seem to have interesting membrane functions and therefore are good candidates to be studied by using the planar lipid bilayer technique.

Incorporation of glycosphingolipids into artificial bilayers has been performed by different authors (Clowes et al., 1971; Deleers et al., 1976; Mkheyan et al., 1981). The presence of some glycolipids (glucosylceramide and galactosylceramide) slightly affects the ionic permeability properties of phospholipid bilayers (Gambale et al., 1982). The ionic conductance increases exponentially as a function of the glycolipid-phospholipid molar ratio. It seems that glycolipids cluster in solid microscopic domains

whose boundary regions determine leaky pathways through which ions
may diffuse almost freely.

Gangliosides have also been incorporated into planar bilayers,
but owing to their very long carbohydrate chains they are very
soluble in water solutions where they give rise to disk-shaped
micelles. It has been demonstrated that these micelles incorporate
into native cell membranes (Gorio et al., 1980) and phospholipid
vesicles (Cestaro et al., 1979), and interact with phospholipid
bilayers (Gambale et al., in preparation). As a probe, we incorporated the ionic modifier Gramicidin D into ganglioside micelles;
when exposed to Gramicidin-rich micelles planar bilayers developed
typical single channel and multichannel current increases which
depend on the membrane lipid content. We also performed capacitance
measurements to obtain information concerning the composition of
bilayers made from ganglioside-phospholipid mixtures and verified
that almost 95% of the gangliosides (GM1, GD1a, GT1) used to prepare
the mixture were still present in the membrane.

As far as interactions between gangliosides and external
ligands are concerned, effects induced by cholera toxin seem relevant. Cholera toxin is a multivalent ligand containing five identical binding subunits (β) and two components (α and γ) which are
the cause of the biological effects of the toxin. Activation of
adenylate cyclase depends on the penetration of α subunits into the
lipid matrix and follows the specific interaction of the toxin with
the monosialoganglioside GM1 (Brady and Fishman, 1979). Local
disturbances and modifications of the membrane organization may
induce increased permeability either to ions or other molecules.

Ionic permeability modifications were observed in planar bilayers formed from GM1-glycerilmonooleate (GMO) mixtures by addition of small amounts (1-70 ng/ml) of cholera toxin (Tosteson, 1978).
Sharp current jumps (conductance $\simeq$ 20 pS, duration $\simeq$ 1 min) greatly
resembling the opening of single channels were recorded. Tosteson
claims that cholera toxin specifically interacts with GM1, because
no current fluctuations were observed when the disialoganglioside
GM2 was used instead of GM1, or in pure GMO membranes.

Some similarities exist between cholera toxin and trophic hormones. They are composed of β binding and α activating subunits
(Brady and Fishman, 1979); moreover the β-chains of gonadotropin,
thyrotropin, and luteinizing hormone are characterized by a segment
of the amino acid sequence which presents significant similarities
to a segment of the β-chain of cholera toxin (Kurosky et al., 1977).
Therefore it has been suggested that cholera toxin binds to the cell
surface by an interaction which mimics that of the hormones.

The presence of glycolipid units in the receptor structure of
thyrotropin (TSH) has been supported by the incorporation of

induce an increase of the ionic permeability owing to deep penetration and movement of the receptor across the bilayer.

REFERENCES

Alvarez, O., Diaz, E. and Latorre, R., 1975, Voltage dependent conductance induced by hemocyanin in black lipid films, Biochim. Biophys. Acta, 389:444.

Bamberg, E., Noda, K., Gross, E. and Lauger, P., 1976, Single-channel parameters of Gramicidin A, B and C, Biochim. Biophys. Acta, 419:223.

Bean, R.C., Shepherd, W.C., Chan, H. and Eichner, J., 1969, Discrete conductance fluctuations in lipid bilayer protein membranes, J. Gen. Physiol., 53:741.

Blumental, R. and Shamoo, A.E., 1979, Incorporation of transport molecules into black lipid membranes, in: "The Receptors," vol. 1, R.D. O'Brien, ed., Plenum Press, New York.

Blumental, R., Klausner, R.D. and Weinstein, J.N., 1980, Voltage-dependent translocation of the asialoglycoprotein receptor across lipid membranes, Nature, 288:333.

Boheim, G., Hanke, W., Barrantes, F.J., Eibl, H., Sackmann, B., Fels, G. and Maelicke, A., 1981, Agonist-activated ionic channels in acetylcholine receptor reconstituted into planar lipid bilayers, Proc. Natl. Acad. Sci. USA, 78:3586.

Brady, R.O. and Fishman, P.H., 1979, Biotransducers of membrane-mediated information, Adv. Enzym., 50:303.

Cartaud, J., Popot, J.L., and Changeaux, J.P., 1980, Light and heavy forms of the acetylcholine receptor from Torpedo marmorata electric organ. FEBS Lett., 121-327.

Cestaro, B., Ippolito, G., Ghidoni, R., Orlando, P. and Tettamanti, G., 1979, Interaction of GM1 ganglioside micelles with multilayer vesicles, Bull. Mol. Biol. Med., 4:240.

Chatelain, P., Deleers, M., Poss, A. and Ruysschaert, J.M., 1979, A specific GT1 ganglioside-luteinizing hormone interaction induces conductance changes in lipid bilayers. Experientia, 35:334.

Clowes, A.W., Cherry, R.J. and Chapman, D., 1971, Physical properties of lecithin-cerebroside bilayers. Biochim. Biophys. Acta, 249-301.

Deleers, M., Poss, A. and Ruysschaert, J.M., 1976, Specific interactions between concanavalin A and glycolipid incorporated into planar bilayer membranes. Bioch. Biophys. Res. Comm., 72:709.

Deleers, M., Chatelain, P., Poss, A. and Ruysschaert, J.M., 1979, Specific interaction between follitropin and GM1 ganglioside incorporated into lipid membranes. Bioch. Biophys. Res. Comm., 89:1102.

Feldberg, S.W. and Delgado, A.B., 1978, Inner voltage clamping. A method for studying interactions among hydrophobic ions in a lipid bilayer, Bioph. J., 21:61.

Fettiplace, R., Gordon, L.G.M., Hladky, S.B., Requena, J.,
    Zingsheim, H.P. and Haydon, D.A., 1975, Techniques in the
    formation and examination of "black" lipid bilayer membranes,
    Chapter 1 in: "Methods in Membrane Biology," vol. 4, E.D.
    Korn, ed., Plenum Press, New York.
Finkelstein, A., Rubin, L.L. and Tzeng, M.C., 1976, Black widow
    spider venom: effect of purified toxin on lipid bilayer mem-
    branes, Science, 193:1009.
Finkelstein, A., and Andersen, O.S., 1981, The gramicidin A channel:
    a review of its permeability characteristics with special re-
    ference to the single-file aspect of transport. J. Membr.
    Biol., 59:155.
Gabler, R., 1978, "Electrical Interactions in Molecular Biophysics,"
    Academic Press, New York.
Gambale, F., Robello, M., Usai, C. and Marchetti, C., 1982, Proper-
    ties of ionic transport through phospholipid-glycolipid arti-
    ficial bilayers, Biochim. Biophys. Acta 693:165
Gomperts, B.D., 1976, Calcium and cell activation, in: "Receptors
    and Recognition," vol. 2, series A, P. Cuatrecasas and M.F.
    Greaves, eds., Chapman and Hall, London.
Goodal, M.C., Bradley, R.J., Saccomanni, G. and Romine, W.O., 1974,
    Quantum conductance changes in lipid bilayer membranes con-
    taining a cholinergic hydrophobic protein from Electrophorus
    electroplax, Biochim. Biophys. Acta, 352:192.
Gorio, A., Carmignoto, G., Facci, L. and Finesso, M., 1980, Motor
    nerve sprouting induced by ganglioside treatment. Possible
    implications for gangliosides on neuronal growth. Brain Res.,
    197:236.
Grollman, E.F., Lee, G., Ambesi-Impiombato, F.S., Meldolesi, M.F.,
    Aloj, S.M., Coon, H.G., Kaback, H.R. and Kohn, L.D., 1977,
    Effects of thyrotropin on the thyroid cell membranes: hyperpo-
    larization induced by hormone-receptor interaction, Proc. Natl.
    Acad. Sci. USA, 74:2352.
Hakomori, S.I., 1981, Glycosphingolipids in cellular interaction,
    differentiation and oncogenesis, Ann Rev. Biochem., 50:733.
Hladky, S.B. and Haydon, D.A., 1972, Ion transfer across lipid
    membranes in the presence of Gramicidin A - I: Studies of the
    unit conductance channel, Biochim. Biophys. Acta, 274:294.
Hughes, R.C., 1975, The complex carbohydrates of mammalian cell
    surfaces and their biological roles, Essays in Biochem., 11:1.
Jones, M.N., 1975, Membrane models, Chapter 8 in: "Biological
    Interfaces," Elsevier, Amsterdam.
Katz, B., 1966, "Nerve, Muscle and Synapsis," McGraw-Hill, New York.
Kemp, G., Dolly, J.A., Barnard, E.A. and Wenner, C.E., 1973, Re-
    constitution of a partially purified end-plate acetylcholine
    receptor preparation in lipid bilayer membranes, Biochem.
    Biophys. Res. Commun., 54:607.
Kolb-Bachofen, V., 1981, Hepatic receptor for asialoglycoproteins.
    Ultrastructural demonstration of ligand-induced micro-aggre-
    gation of receptors, Biochim. Biophys. Acta, 645:293.

Kurosky, A., Markel, D.E., Peterson, J.W. and Fitch, W.M., 1977, Primary structure of cholera toxin β-chain: a glycoprotein hormone analog, Science, 195:299.

Ledeen, R.W. and Yu, R.K., 1973, Structure and enzymic degradation of sphingolipids, Chapter 4 in: "Lysosomes and Storage Diseases," H.G. Hers and F. Van Hoof, eds., Academic Press, New York.

Lindstrom, J., Anholt, R., Einarson, B., Engel, A., Osame, M. and Montal, M., 1980, Purification of acetylcholine receptors, reconstitution into lipid vesicles and study of agonist-induced cation channel regulation, J. Biol. Chem., 255:8340.

McLaughlin, S., 1977, Electrostatic potentials at membrane-solution interfaces, in: "Current Topics in Membranes and Transport," vol. 9, Academic Press, New York.

Menestrina, G. and Antolini, R., 1982, The dependence of the conductance of the hemocyanin channel on applied potential and ionic concentration with mono- and divalent cations, Biochim. Biophys. Acta, 668:673.

Mkheyan, E.E., Akopov, S.E., Sotsky, O.P. and Badzhinyan, S.A., 1981, Some physical parameters of cerebroside and phospholipid-cerebroside membranes, Biofizika, 26:48.

Montal, M. and Mueller, P., 1972, Formation of bimolecular membranes from lipid monolayers and a study of their electrical properties, Proc. Natl. Acad. Sci. USA, 69:3561.

Montal, M., Darzson, A. and Schindler, H., 1981, Functional reassembly of membrane proteins in planar lipid bilayers, Q. Rev. Biophys., 14:1.

Mueller, P., Rudin, D.O., Ti Tien, H. and Wescott, W.C., 1962, Reconstitution of excitable membrane structure in vitro, Circulation, 26:1167.

Mullin, B.R., Fishman, P.H., Lee, G., Aloj, S.M., Ledley, F.D., Winaud, R.J., Kohn, L. and Brady, R.O., 1976, Thyrotropin-ganglioside interactions and their relationship to the structure and function of thyrotropin receptors, Proc. Natl. Acad. Sci. USA, 73:842.

Neher, E. and Sackmann, B., 1976, Single-channel currents recorded from membrane of denervated frog muscle fibres, Nature, 250:68.

Nelson, N., Anholt, R., Lindstrom, J. and Montal, M., 1980, Reconstitution of purified acetylcholine receptors with functional ion channels in planar lipid bilayers, Proc. Natl. Acad. Sci. USA, 77:3057.

O'Keefe, E. and Cuatrecasas, P., 1978, Cholera toxin and membrane gangliosides: binding and adenylate cyclase activation in normal and transformed cells, J. Membrane Biol., 42:61.

Parisi, M., Rivas, E. and De Robertis, E., 1971, Conductance changes produced by acetylcholine in lipidic membranes containing a proteolipid from Electrophorus, Science, 172:56.

Poss, A., Deleers, M. and Ruysschaert, J.M., 1978, Evidence for a specific interaction between GT1 ganglioside incorporated into bilayer membranes and thyrotropin, FEBS Lett., 86:160.

Rasmussen, H., 1975, Ions as "second messengers," in: "Cell Mem-
    branes. Biochemistry, Cell Biology and Pathology," G. Weiss-
    mann and S. Claiborne, eds., HP Publishing Company, New York.
Reader, T.A. and De Robertis, E., 1974, The response of artificial
    lipid membranes containing a cholinergic hydrophobic protein
    from Electrophorus electroplax, Biochim. Biophys. Acta, 352:
    192.
Schindler, H. and Rosenbush, J.P., 1978, Matrix protein from
    Escherichia coli outer membranes forms voltage-controlled
    channels in lipid bilayers, Proc. Natl. Acad. Sci. USA, 75:
    3751.
Schindler, H. and Quast, U., 1980, Functional acetylcholine recep-
    tor from Torpedo marmorata in planar membranes, Proc. Natl.
    Acad. Sci. USA, 77:3052.
Szabo, G., Eisenman, G., Laprade, R., Ciani, S.M. and Krasne, S.,
    1973, Experimentally observed effects of carriers on the
    electrical properties of bilayer membranes-equilibrium domain,
    in: "Membranes," vol. 2, G. Eisenman, ed., Marcel Dekker,
    New York.
Takagi, M., Azuma, K. and Kishimoto, U., 1965, A new method for the
    formation of bilayer membranes in aqueous solution, Annu. Rep.
    Biol. Works Fac. Sci. Osaka Univ., 13:107.
Ti Tien, H., 1974, "Bilayer lipid membranes (BLM). Theory and
    practice," Marcel Dekker, New York.
Tosteson, M.T. and Tosteson, D.C., 1978, Bilayers containing
    gangliosides develop channels when exposed to cholera toxin,
    Nature, 275:142.
Urry, D.W., 1971, The gramicidin A trasmembrane channel: a proposed
    π(L.D) helix, Proc. Natl. Acad. Sci. USA, 68:672.
Vodyanoy, V. and Murphy, R.B., 1982, Solvent-free lipid bimolecular
    membranes of large surface area, Biochim. Biophys. Acta,
    687:189.
White, S.H., 1978, Formation of "solvent-free" black lipid bilayer
    membranes from glyceril monooleate dispersed in squalene,
    Biophys. J., 23:337.
Wu, W.C.S., Moore, H.H. and Raftery, M.A., 1981, Quantitation of
    cation transport by reconstituted membrane vesicles con-
    taining purified acetylcholine receptor, Proc. Natl. Acad.
    Sci. USA, 78:775.

STRUCTURE-FUNCTION RELATIONS FOR CELL SURFACE RECEPTORS
AND ADENYLATE CYCLASE: STUDIES USING TARGET SIZE ANALYSIS
OF RADIATION INACTIVATION

Werner Schlegel
Fondation pour recherches medicales
Dept. of Medicine, University of Geneva
64, av. de la Roserie, CH-1211 Geneva 4, Switzerland

INTRODUCTION : METHODS FOR THE STUDY OF STRUCTURE-
FUNCTION RELATIONS FOR CELL SURFACE RECEPTORS :

Understanding the functioning of a cell surface receptor is possible only with knowledge about its structure. A priori receptors are likely to be complex since
many functions have to be accomodated. A receptor reco-
gnizes a particular substance or class of substances with
great precision in a sea of similar substances, binds it,
generates a signal by triggering contacting enzyme systems or by regulation of ion channels, terminates the
signal, and is subjected to various modulations by "non-
specific" ligands like GTP, ions, and phospholipids, or
by the specific ligands, i. e. in the instance of desensitization. In a first approach to structural analysis
the question is asked what structure is responsible for
a particular partial function of the receptor. Several
methods have been applied to the study of structure-
function relationships of cell surface receptors :

Affinity Labelling

The first step in receptor function is the binding
of the ligands. The structures in contact with the ligand can be tagged either by using a modified ligand
that can be activated (i.e. by photons) to chemically
react with the receptor, or by the use of a crosslinking
chemical. The macromolecules identified in this manner

233

are those that have the closest and longest lasting contact with the ligand. Note that not all parts of a receptor necessary for binding may be labelled. In addition it is necessary to demonstrate that a ligand that
is crosslinked to the receptor exerts its action as an
agonist or antagonist, in order to prove that the relevant site has been labelled.

Purification

The major challenge in the purification of cell surface receptors is their organization in the membrane; the
crucial step thus is the solubilization. Conditions have
to be found that maintain the basic structural organization, yet lead to real solubilization, separating the receptor from other membrane elements. High affinity binding and ligand specificity in the solubilized state is
the usual criterion for the intactness of receptor structure. An even better way to check for functionality is
to prove the capacity of a solubilized receptor to couple
to the effector system. This has been done in an elegant
way for the beta receptor for catecholamines[1] . For the
purification one can take advantage of the specific properties of a particular receptor such as ligand binding
(affinity chromatography) antigenicity (purification with
specific antibodies), or glycosylation (lectin columns).

Reconstitution

The function of a receptor component is most clearly
demonstrated when it can reconstitute a system that lacks
this function. By fusion of cells or membranes carrying
different types of receptors it was demonstrated that
elements of receptor systems are interchangeable[2] . This
has opened the possibility to combine elements of different receptor systems to form new functional assemblies.
The problem is to find or create receptor systems in
which a particular well defined element is missing or
non-functional. For the adenylate cyclase system a mutant
was described that is deficient in the coupling of the
receptor to the catalytic moiety[3] . Using the reconstitution of this system as an assay, the stimulatory nucleotide regulatory unit was purified[4,5] . It is however
still not clear what crucial structure is really missing
or defective in the mutant. Since the process of reconstitution for which very particular experimental conditions

are required [4,5] , is not well understood. It is conceivable that parts of the purified substructures are required
only for the reassembly of a functional adenylate cyclase
system and may not per se participate in the coupling of
the receptor to the catalytic moiety. Even more problematic is the reconstitution of receptor systems that are
partially inactivated by protein modification reagents,
since the nature of the damage cannot be clearly defined,
and thus the function of proteins that can "repair" such
damage remains ambiguous. The ultimate goal of reconstitution experiments is to reassemble a functional receptor
system from purified elements. This has so far only be
achieved for the acetylcholine receptor, a receptor that
can be purified in large quantities from specialized organs, and is the result of intensive efforts over a long
period of time (see the chapter by J. P. Changeux in this
book).

Indirect Method

By measuring the kinetics of various functions of a
receptor system such as binding, stimulation or inhibition
of adenylate cyclase or other effector enzymes, or modulation of binding or coupling by GTP or metal ions, data
is created that reflects the structural organization of
the various elements of a receptor system. The difficulty
is that the date has often a degree of complexity that
makes modelling of the date a futile effort, since the
elevated number of model parameters does no longer allow
to distinguish between very different models, and hence
the number of models proposed i.e. for adenylate cyclase
systems is too vast or limit the imagination on a possible structural organization of the system.

Target Size Analysis

It is possible to probe cell surface receptor structure in the intact membrane by radiation inactivation, a
method that allows to determine the functional size of
an enzyme or receptor without the need for purification
or solubilization. Theory, potential and problematics of
the application of this method will be discussed in detail
below. Aside from electron microscopy this seems the only
method that directly probes for structure of the receptor
in situ. It can therefore make a valuable contribution
to the study of structure-function relationships.

TARGET SIZE ANALYSIS OF CELL SURFACE RECEPTORS AND
ADENYLATE CYCLASE

Theory

A very general outline of the method is straightforward. Biological samples are irradiated under conditions that avoid indirect effects of radiation. The surviving activity of an enzyme or receptor under study is
related to the radiation dose delivered to the sample
and the inactivation as a function of dose permits the
determination of the target size.

The explanation of radiation effects in terms of a
target analysis was developed by Lea[6] . It incorporates
several general principles which are briefly described
here. Ionizing radiation passes through free space una-
bated; when it passes through matter, interactions occur
in which large amounts of energy are transferred to the
atom which is hit. This process, termed "primary ionization", involves an average energy transfer of 65 electron volts (about 1500 kcal/mol) which is dissipated in
the target molecule. Covalent bond breakage, excitations
and a variety of other processes result. Severe structural damage occurs in the molecule and biological activity is destroyed. Other molecules which did not suffer a
primary ionization are undamaged and retain full biological activity. If the radiation causes primary ionizations
at random along its trajectory through matter (as is the
case for γ-rays and high energy electrons), the probability of n primary ionizations in a given unit follows a
Poisson distribution

$$P(n) = \frac{e^{-x}.x^n}{n!}$$

where x is the average number of primary ionizations
occuring per unit. Since biological activity will only
be seen in molecules which have escaped damage, it is of
interest to determine the probability of escape i.e.,
no primary ionizations : since $x^0 = 1$ and $0! = 1$

$$P(0) = e^{-x}$$

The average number of primary ionizations per unit, x,
is given by kD where D is the dose of radiation and k is
a constant proportional to the size of the unit. The
amount of biological activity which survives irradiation
is directly proportional to the fraction of biologically
active units which have escaped radiation damage : Thus

$$\frac{A}{A_O} = e^{-x} = e^{-kD}$$

where A is the surviving activity of the sample submitted
to the radiation dose D, and A_O is the activity of the
unirradiated control.
Thus the size of the functional unit required for the
enzymatic activity A is proportional to the exponential
coefficient k of the inactivation curve. One can relate
k to the mass of the functional unit i.e. expressed in
daltons; if the radiation dose D would be expressed as
the number of primary ionizations per dalton, k would
represent the mass in daltons. Since radiation dose is
normally expressed in rads, a proportionality factor has
to be introduced :

1 rad = 100 ergs/gram = 6.24×10^{13} eV/gram

$\qquad$ = 9.6×10^{11} ionizations/gram = 1.59×10^{-12} ion./dalton

$k(\text{dalton}) = k\left(\frac{1}{\text{rad}}\right) \times \frac{1}{1.59 \times 10^{-12}} = k\left(\frac{1}{\text{rad}}\right) \times 6.3 \times 10^{11}$

An empirical proportionality factor of 6.4×10^{11} was reported by Kepner and Macey . Frequently $D_{37} = 1/k$, i.e.
the dose reducing activity to $0.37 \times A_O$ is used in the
calculations (ln (0.37) = -1;

for $A_D/A_O = 0.37$ ln $(A_D/A_O) = -1$

and thus $-kD_{37} = -1$ i.e. $D_{37} = 1/k$).

The Irradiation Inactivation Experiment

Radiation source. High energy electrons or gamma
rays are the forms of ionizing radiation used for molecular weight determination. High energy electrons at sufficiently high dose rates can be obtained from van der
Graaf or some linear accelerators. The energy of the
electrons determines their penetration power. This has
to be considered in design of the experimental setup; in
order to cause random ionizations throughout the sample,
electrons must fully traverse the material.
Dosimetry. Experimental conditions must be arranged
to insure that the radiation dose is uniformly delivered
over all samples which are simultaneously exposed and
also that the dose can be accurately determined. It is
advisable to use known enzymes as internal controls in
the irradiated samples. Both enzymes inherent in the

experimental preparation as well as exogenous enzymes
which were added to the samples before freezing have been
used. The radiation inactivation of these activities offe[
not only an independent estimate of the radiation dose,
but also indicate variations in the samples due to other
factors.

Sample preparation and handling during irradiation.
Target theory is invalid for radiation exposure of samples in the liquid state[6,8] . Lyophilized samples and
frozen preparations have been shown to be appropriate materials for this technique. For irradiations above freezing temperatures it is necessary to remove all volatile
water and take the sample to dryness. Extremely low concentrations of proteins - 1 µg/ml or less - could result
in spurious loss of activity due to binding of the sample
by charges produced by radiation on glass vials. Protein
which is inactive in the biological assay of the sample
can be added as a carrier to raise the protein level.
Samples to be irradiated as frozen solutions have to be
maintained at low temperature throughout the inactivation
experiment, and if they are sealed in air, ozone formed
during irradiation should be allowed to escape prior to
thawing. Since the apparent target size is a function of
the temperature during exposure to radiation a correction
factor has to be introduced in the final calculation[8,9,10]
and it is necessary to maintain the samples at constant
temperature during irradiation.
If the lyophilization or freezing of the sample affects
the activity, application of the method becomes problematic since target sizes eventually determined will only
apply for the fraction of the activity surviving procedures involved in sample preparation.
Assay. Inactivation by radiation is an exponential
function of dose; as a consequence a sufficient range of
inactivation has to be covered in order to recognize wether the decay curve is monoexponential or more complex.
It is therefore necessary that the assay allows to measure with good accuracy surviving activity of a few percent of the original activity. A note of caution applies
in particular to binding assays: Since specific binding
of a ligand to a cell surface receptor is normally determined as the difference between total and nonspecific
binding and the latter does not decrease with radiation,
the errors in the difference increase with increasing
inactivation; this may for some receptor systems preclude
the use of the technique.

Interpretation of Radiation Inactivation Data

There is a fundamental difference between target
size analysis and other methods used to determine molecular weights. Classical methods such as gelelectrophoresis, gelchromatography, or sedimentation analysis determine the physical state of the macromolecule that is
investigated, whereas with target size analysis one (or
several) functions(s) of a molecular assembly is assayed,
the size thus obtained representing a functional unit,
i.e. the sum of the masses of all macromolecules which
are required in the assay (see below). The potential of
the method, but also the major problems in its application relate to this basic difference. For membrane bound
proteins the physical state of the macromolecular assembly that is investigated is obviously a function of the
solubilization conditions that are employed, whereas target sizes can be obtained for both the membrane bound
and the solubilized states. The potential of the method
thus lies in the fact that it probes for structure with
minimal perturbation of the systems and allows i.e. com-
parisons between membrane bound and solubilized states.
The size of a given structure is however always infered
indirectly and the analysis relies on some basic assumptions which cannot be verified in all instances.

The functional unit concept. Reviewing radiation
inactivation data on a wide variety of enzymes with
known molecular size and subunit composition we found
that the target size could correspond to either the size
of subunits, subunits assemblies, or that of the whole
molecule[9]. The first possibility to explain this finding
is that the primary ionization destroys a single subunit;
if the subunit is fully independent the remaining subunits in a complex remain functional and the target size
obtained is that of the subunit; if however the interaction between different subunits is required for the
measured activity the loss of any subunit in a complex
would inactivate the complex and in that case the target
size observed would be that of the minimal assembly of
subunits required for a given function, i.e. the size of
the functional unit[9]. It should be possible for a macromolecular complex with several independent functions to
find different target sizes for the different functions;[11]
this is indeed the case for multienzymes clusters as
well as for the adenylate cyclase system (see below).

Energy transfer. An alternative explanation for the
fact that the target size corresponds to either the size

of subunits or of subunit assemblies is to assume that
sufficient energy could be transferred between adjacent
subunits to destroy a subunit assembly with one primary
ionization. Wether individual subunits would be destroyed
or whole assemblies would reflect the closeness of association of these subunits. So far there is no direct evidence to support this idea[9]; in fact there seems to be in-
sufficient energy transfer between the covalently attached carbohydrate moiety of a glycoprotein and the protein part carrying the catalytic function to lead to a
difference in target sizes between glycosylated and non-
glycosylated invertase[12]. It is however possible that the
damage to the assembly does not arise during irradiation
and by radiation per se but during the process of rehy-
dration or other changes occuring between irradiation
and assay. A radiation damaged subunit of a multisubunit
protein could prevent the reformation of an active form
of the whole assembly (destruction transfer). Such a
phenomenon would be of particular importance if the physical state of the enzyme as it is irradiated is very
different from the state during the assay and it could
explain inconsistencies in target sizes obtained when an
enzyme is irradiated lyophilized from different buffers[13]
or as a precipitate when compared to the inactivation of
a frozen solution[11].

    Modelling of Target Data. The simple mathematics of
target theory makes modelling of target data an easy task.
An instructive example was given by Kincaid et al.[14] for
data on calmodulin dependent phosphodiesterase. Modelling
can demonstrate that in a complex equilibrium system misleading conclusions can be reached when simple target
analysis is uncritically applied[15].

    In summary interpretation of radiation inactivation
is always possible in several ways. It is however clear
that a part of the interpretation will not change wether
i.e. energy transfer (or destruction transfer) is or is
not considered. Computer modelling of target data can be
of great help to appreciate the limitations of the target
sizing approach and should allow on to choose selective
experimentation. If molecular sizes obtained with other
methods can be used for modelling the informational value of radiation inactivation data will become much larger

THE CONTRIBUTION OF TARGET ANALYSIS TO THE UNDERSTANDING
OF STRUCTURE-FUNCTION RELATIONS FOR CELL SURFACE
RECEPTORS AND ADENYLATE CYCLASE

Target size analysis probes for the structure of
cell surface receptors and adenylate cyclase systems as
they exist in the intact membrane and can offer predic-
tions on the size of a functional receptor assembly. As
was shown for the insulin receptor (see below) radiation
inactivation can reveal novel aspects in the receptor
organization that did not show in any other analysis.
The complexity of receptor systems and our current state
of relative ignorance about the relevant biochemical
events in signal transmission limits the interpretation
of target data on receptors and especially adenylate cyclase systems. This is no reason to refrain from the use
of this powerful technique, but it is advisable to keep
an open mind and accept several possibilities of interpretation of the same target data. The following brief
summary of some major results from the application of the
technique cannot include all possibilities of interpretation, and for a comprehensive view the reader is referred to the original publications.

Insulin Receptors

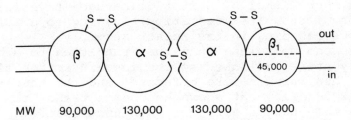

Scheme 1 : Proposed subunit structure of the insulin
receptor [16] . The receptor structures were identified
by affinity labelling or purification and the subunit
sizes were determined on SDS gelelectrophoresis.

Based mainly on affinity labelling studies the structure depicted in scheme 1 has been proposed for the insulin receptor [16]. Target size analysis of the insulin receptor has been carried out in three systems: rat liver plasma membranes [17,18] and solubilized membranes of the lymphoblasma cell IM-9 [19]. The results show striking differences (Fig.1). The solubilized IM-9 receptor shows a target size of 170,000 MW, consistent with an assembly of 2 subunits (120,000 and 50,000 MW) being required for insulin binding. The inactivation data follows a single exponential decay and since linear Scatchard plots are obtained target size analysis is straightforward. With rat liver plasma membranes more complex data is obtained, even for solubilized preparations. Low doses of radiation lead to an increased affinity of the insulin receptor[17] and as a consequence to an increase in insulin binding at low insulin concentration. The inactivation data reveals that a structure of approximately 350,000 MW reduces the affinity of the insulin binding sites, and a structure of about 90,000 MW is responsible for the high affinity binding, after elimination of the larger structure by irradiation [17]. If the receptor is shifted to the high affinity state[21,22] by either insulin trypsin, or changes in pH the influence of the larger structure is partially or completely abolished[18,20] . These results indicate that the complete receptor assembly prevents the transition to the high affinity state of the receptor, whereas subunits as small as 90,000 MW are independently capable of binding insulin with high affinity. Transition to the high affinity state can be triggered by trypsin or low doses of radiation,thus either by partial cleavage (trypsin) or elimination of subunits (radiation) leading to a loss of the functions residing in the intact assembly.

One can speculate that the transition to a high affinity state is linked to the insulin action since trypsin has insulin-like effects [16] and the insulin receptor in IM9-cells that shows neither a transition[23] nor the assembly target size[19] is nonfunctional.

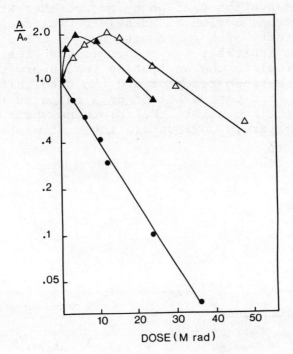

Fig. 1. Comparison between radiation inactivation of insulin binding in rat liver plasma membranes,
native (△), and solubilized (▲), versus in
solubilized membranes from lymphoma cells,
IM-9 (●). Data was replotted[17,19] and represents
either binding of ^{125}I-insulin at tracer concentration[17], or insulin binding capacity derived
from Scatchard plots [19] as a function of increasing radiation dose.

Adenylate Cyclase

There is general agreement that adenylate cyclase systems are composed of three basic units: the catalytic moiety, the nucleotide regulatory proteins, and the receptor (see scheme 2). Note that the concept originated from the observation that guanylnucleotides are essential for hormone action on adenylate cyclase and in - fluence affinity of hormone binding[24,25] . The easiest but still an arbitrary way to accomodate these features was to intercalate the nucleotide regulatory units as transducers between receptor and the catalytic moiety. Despite the fact that a lot of data supports this notion now, it is still possible that proteins whose functions are not immediately obvious are integral parts of the systems.

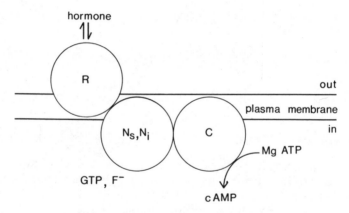

Scheme· 2 : Postulated general scheme for the structure of adenylate cyclase : R : receptor; N_s, N_i : nucleotide regulatory units that mediate the action of GTP on receptor binding, GTP and F^- action on adenylate cyclase, and coupling of receptors to the catalytic moiety, N_s mediate coupling of stimulatory, N_i of inhibitory receptors; C : catalytic moiety.

The catalytic moiety, C. Adenylate cyclase can exist in a state where regulation by hormones, guanylnucleotides, and F^- is lost and only Mn-ATP can serve as a

substrate. It is thought that these properties reflect
the isolated catalytic moiety C, since they arise from
fractionation of solubilized enzymes on gel filtration
 26,27 and exist in ram sperm [28] and in mutants of lymphoma cells [3] (cyc⁻) that are defective in regulation of
catalytic activity by hormones, guanylnucleotides, and
F⁻ [29]. Proteolytic treatment of these membranes that are
devoid of a functional nucleotide regulatory component
leads to the solubilization of a 46,000 MW protein that
has adenylate cyclase activity only with Mn-ATP as substrate [30]. Target analysis of basal adenylate cyclase activity with Mn-ATP has yielded target sizes of 92,000 MW
and 150,000 MW for turkey erythrocyte [31] and liver[9]
respectively (table 2); the latter size being probably
overestimated due to a nonlinear decay curve[9]. These are
the smallest target sizes observed in adenylate cyclase
systems. Target size analysis thus predicts that the size of C as it exists in membranes containing functional
nucleotide regulatory units is considerably larger than
the proteolytic fragment. Note that in those membranes C
cannot be solubilized by proteolysis [30].

Table 1.   Target Sizes for Distinct Functions of the
           Adenylate Cyclase System.

C: Catalytic activity expressed with Mn^{2+} in the
              absence of activators :

              rat liver :   150,000 (ref. 9)
              turkey ec :   92,000 (ref. 31)

N_SC: Adenylate cyclase activity stimulated by F⁻,
              or GppNHp, after preactivation :

              rat liver   :   230,000 (9) 220,000 (32)
              rat fat cells :   220,000 (33)
              turkey ec   :   250,000 (31)

RN_SC: Adenylate cyclase activity stimulated by hor-
              mone and GTP after preactivation :

              rat liver   :   350,000 (9)

              Activation of adenylate cyclase by hormone
              and GppNHp :

              turkey ec   :   316,000 (31)

The nucleotide regulatory units. The precise biochemical mechanism by which these regulatory proteins convey
the changes in catalytic activity of C due to guanylnucleotides or fluoride are still unknown. Therefore in any
analysis they are arbitrarily defined i.e. as the proteins
that can reconstitute a defective system, or also by the
fact that they are ADP-ribosylated by cholera toxin and
can be affinity labelled with analogues of GTP. Target
analysis can only determine the size of N_sC which is very
similar for liver[9,32], turkey erythrocyte[31]and fat cell[33]
cyclase. Note that the size for N_sC in liver and fat cell
adenylate cyclase is observed after preactivation; the activation process itself depends on a different functional
unit (see below). One can speculatively refer to the increment between C and N_sC as the target size of N which
would then be between $100,000$[4,5] and 130,000 MW. Compared
to the subunit sizes of N_S defined by other analysis of
35,000, 42,000, 45,000 and 53,000 MW, the large increment
indicates that more than a single subunit is required for
a functional N_S.

Receptors in adenylate cyclase systems. So far target analysis has been done for the beta-receptor in turkey
erythrocyte[32] and the glucagon-receptor in rat liver membranes[9]. For both an increment of about 100,000 MW is observed between NC and RNC target sizes. In rat liver membranes a protein of 53,000 MW could be affinity labelled
with glucagon[34]; for the beta-receptor for catecholamines
subunit sizes of 37,000 and 41,000 MW[35] and 32,000 MW[36]
have been reported for turkey erythrocytes and other systems. This indicates a minimum dimer assembly of subunits
required for both receptors in their activatory function
in adenylate cyclase. Whereas target sizes for binding
and activation in turkey erythrocytes membranes match[31],
the target size for GTP displaceable glucagon binding in
rat liver membranes is much larger than the difference
between RNC and NC[9].

Structural changes during the activation process.
Radiation inactivation can also be carried out prior to
exposure of the plasma membranes to any activating ligand,
i.e. in the ground state of the system. If the target sizes in the activated states are compared with those obtained for the ground state of the enzyme, the functional
unit sizes for the activation process can be obtained. Radiation inactivation curves for the ground state enzyme
from liver and fat cell membranes are nonlinear (Fig.2).
Note that the nonlinearity can only be appreciated in the
lower half of the curve, hence the importance of
carrying the inactivation to low surviving

activities. Nonlinear decay curves are indicative of multiple targets, i.e. more than a single species is involved in the observed function; their interpretation requires the application of a model. For adenylate cyclase
there is a vast choice of possible models developed from
kinetic data and all the models have a high degree of
complexity; therefore we chose to apply the simplest
possible model for the interpretation of this data, which
resulted in the proposition that a large target size MW
in the range of 10^6 was responsible for activation[9,33].
Interestingly, nonlinear decay curves were obtained
independently of the ligand (GTP, F^-, GppNHp, glucagon)
used for the activation in the assay. This means that
even for ligands which are thought to act directly
through the nucleotide regulatory unit (Scheme 1) F^- or
GppNHp, the activation process depends on more than a
single molecular species. Radiation inactivation curves
for turkey erythrocyte adenylate cyclase[31] are linear
for both the ground state as well as the preactivated
states of the enzyme (Fig.2). This can indicate that
there is a fundamental difference in the structure between these systems; alternatively the process of activation could rely on different biochemical events. An alternative approach was taken by Martin and coworkers[32],
assuming energy transfer between the different parts of
the adenylate cyclase in the "coupled" states . As a consequence the experimentation that was chosen is mostly
very different from that in our work. In addition, the
membrane preparation was irradiated in the lyophilized
state, which in our own experience leads to marked changes in the kinetic properties of the glucagon stimulated
adenylate cyclase. While it is difficult to compare the
different data, the target size proposed for N_sC is
matched[9,32] (Table 1). The discrepancies in the conclusions
about structural changes during the activation process
between the work by Martin and our own are difficult to
understand. They are however no novelty in the adenylate
cyclase field. Based on the kinetic analysis of the system some diametrically opposed models were proposed.
Thus the analysis with any approach reflects the lack of
easily measurable clearly defined biochemical events
involved in the regulation of the adenylate cyclase system. None of the approaches has so far come to definitive
conclusions about structure-function relationships in
the adenylate cyclase system. Target size analysis has
pointed out some unexpected possibilities for the structural organization of the system. Those, obviously, will

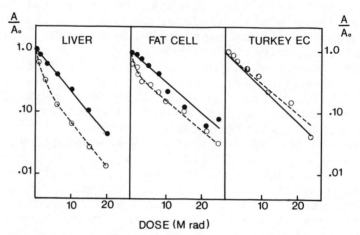

Fig. 2. Comparison between radiation inactivation of
adenylate cyclase in liver, fat cell, and
turkey erythrocyte membranes irradiated in the
ground state (○) and assayed either with
GppNHp for liver and fat cell, or F⁻ for turkey
erythrocyte membranes. Radiation inactivation
for the enzyme pretreated with GppNHp for liver
and fat cell (●) or with isoproterenol plus
GMP for the turkey erythrocyte membranes and
assayed in the presence of GppNHp (solid line)
is shown for reference. For details we refer to
ref. 9 (liver), ref. 33 (fat cell) and ref. 31
(turkey erythrocyte).

have to be verified with other approaches.

CONCLUSIONS

Structural analysis of cell surface receptors and receptor linked enzyme systems has in recent years added new insights into the molecular biology of receptor function. Due to the complexity of receptor structure and function only the combination of all possible approaches leads to a progress in the field. Among the techniques used for probing cell surface receptor structure radiation inactivation has the unique advantage to allow the study of a receptor in its membrane environment. In addition to the examples mentioned above the approach has been applied to receptor for low density lipoproteins [37] asialoglycoproteins[38], immunoglobulin E[39], benzodiapezine[40,41,42] and acetylcholine[43,44] . Despite the limitations in the interpretation discussed in this paper it can make a valuable contribution to our understanding of structure-function relationships in cell surface receptors.

Acknowledgments : I am indebted to Dr. E. S. Kempner for many helpful and stimulating discussions. Financial support by the Fonds National Suisse de la Recherche Scientifique is greatfully acknowledged (grant Nr. 3.933-.80).

REFERENCES

1. Y. Citri and M. Schramm, Resolution, reconstitution and kinetics of the primary action of a hormone receptor, Nature 287:297 (1980).
2. J. Orly and M. Schramm, Coupling of catecholamine receptor from one cell with adenylate cyclase from another cell by cell fusion, Proc. Nat. Acad. Sci. 73:4410 (1976).
3. H. R. Bourne, P. Coffino, and G.M. Tomkins, Selection of a variant lymphoma cell deficient in adenylate cyclase, Science 187:750 (1975).
4. P. C. Sternweis, J. K. Northup, M. D. Smigel, and A. G. Gilman, The regulatory component of adenylate cyclase. Purification and properties, J. Biol. Chem. 256:11517 (1981).

5. E. Hanski, P. C. Sternweis, J. K. Northup, A. W. Dromerick, and A. G. Gilman, The regulatory component of adenylate cyclase. Purification and properties of the turkey erythrocyte protein, J. Biol. Chem. 256:12911 (1981).

6. D. E. Lea, Actions of radiation on living cells, Cambridge University Press, London (1955).

7. G. R. Kepner and R. I. Macey, Membrane enzyme systems. Molecular size determinations by radiation inactivation, Biochim. Biophys. Acta 163:188 (1968).

8. E. S. Kempner and W. Schlegel, Size determination of enzymes by radiation inactivation, Anal. Biochem. 92:2 (1979).

9. W. Schlegel, E. S. Kempner, and M. Rodbell, Activation of adenylate cyclase in hepatic membranes involves interactions of the catalytic unit with multimeric complexes of regulatory proteins, J. Biol. Chem. 254:5168 (1979).

10. E. S. Kempner and H. T. Haigler, The influence of low temperature on the radiation sensitivity of enzymes, J. Biol. Chem., in press (1983).

11. E. S. Kempner, J. H. Miller, W. Schlegel, and J. Z. Hearon, The functional unit of polyenzymes. Determination by radiation inactivation, J. Biol. Chem. 255:6826 (1980).

12. M. E. Lowe and E. S. Kempner, Radiation inactivation of the glycoprotein, invertase, J. Biol. Chem., in press (1982).

13. D. Parkinson and B. A. Callingham, Irradiation inactivation analysis of acetylcholinesterase and the effect of buffer salts, Rad. Res. 90:252 (1982).

14. R. L. Kincaid, E. Kempner, V. C. Manganiello, J. C. Osborne Jr., and M. Vaughan, Calmodulin-activated cyclic nucleotide phosphodiesterase from brain. Relationship of subunit structure to activity assessed by radiation inactivation, J. Biol. Chem. 256 (21):11351 (1981).

15. P. Simon, S. Swillens, and J. E. Dumont, Size determination of an equilibrium enzymic system by radiation inactivation, theoretical considerations, Biochem. J. 205:477 (1982).

16. M. P. Czech, J. Massague, and P. F. Pilch, The insulin receptor : structural features, TIBS 68:222 (1981).

17. J. T. Harmon, C. R. Kahn, E. S. Kempner, and W. Schlegel, Characterization of the insulin receptor in its membrane environment by radiation inactivation, J. Biol. Chem. 255:3412 (1980).

18. J. T. Harmon, E. S. Kempner, and C. R. Kahn, Demonstration by radiation inactivation that insulin alters the structure of the insulin receptor in rat liver membranes, J. Biol. Chem. 256:7719 (1981).

19. R. J. Pollet, E. S. Kempner, M. L. Standaert, and B. A. Haase, Structure of the insulin receptor of the cultured human lymphoblastoid cell IM-9, evidence suggesting that two subunits are required for insulin binding, J. Biol. Chem. 257:894 (1982).

20. J. T. Harmon and J. A. Hedo, Characterization of the chemical and functional nature of a membrane regulator of insulin receptor affinity, 63rd meeting endocrine society (Cincinatti), abstract No 259: 147 (1981).

21. R. E. Corin and D. B. Donner, Insulin receptors convert to a higher affinity state subsequent to hormone binding, a two-state model for the insulin receptor, J. Biol. Chem. 257:104 (1982).

22. D. B. Donner and R.E. Corin, Formation of a receptor state from which insulin dissociates slowly in hepatic cells and plasma membranes, J. Biol. Chem. 255:9005 (1980).

23. R. J. Pollet, M. L. Standaert, and B.A. Haase, Insulin binding to the human lymphocyte receptor, evaluation of the negative cooperativity model, J. Biol. Chem. 252:5828, (1977).

24. M. Rodbell, The role of hormone receptors and GTP-regulatory proteins in membrane transduction, Nature 284:17 (1980).

25. A. M. Spiegel and R. W. Downs, Jr, Guanine nucleotides : key regulators of hormone receptor-adenylate cyclase interaction, Endocrine Reviews 2:275 (1981).

26. Th. Pfeuffer, Guanine nucleotide-controlled interactions between components of adenylate cyclase, FEBS Letts 101:85 (1979).

27. S. Strittmatter and E. J. Neer, Properties of the separated catalytic and regulatory units of brain adenylate cyclase, Proc. Natl. Acad. Sci. 77:6344 (1980).

28. D. Stengel and J. Hanoune, The catalytic unit of ram sperm adenylate cyclase can be activated through the guanine nucleotide regulatory component and prostaglandin receptors of human erythrocyte, J. Biol. Chem. 256:5394 (1981).

29. E. M. Ross, A. C. Howlett, K. M. Ferguson, and A. G. Gilman, Reconstitution of hormone-sensitive adenylate cyclase activity with resolved components of

the enzyme, J. Biol. Chem. 253:6401 (1978).

30. D. Stengel, L. Guenet, and J. Hanoune, Proteolytic
    solubilization of adenylate cyclase from membranes
    deficient in regulatory component, J. Biol. Chem.
    257:10818 (1982).

31. T. B. Nielsen, P. M. Lad, M. S. Preston, E. Kempner,
    W. Schlegel, and M. Rodbell, Structure of the turkey
    erythrocyte adenylate cyclase system, Proc. Natl.
    Acad. Sci. 78:722 (1981).

32. R. B. Martin, J. M. Stein, E. L. Kennedy, C. A.
    Doberska, and J. C. Metcalfe, Transient complexes,
    a new structural model for the activation of adeny-
    late cyclase by hormone receptors (guanine nucleo-
    tides/irradiation inactivation), Biochem. J. 184:
    253 (1979).

33. W. Schlegel, D. M. F. Cooper, and M. Rodbell, Inhibi-
    tion and activation of fat cell adenylate cyclase
    by GTP is mediated by structures of different size,
    Arch. Biochem. Biophys. 201:678 (1980).

34. G. L. Johnson, V. I. MacAndrew Jr, and P. F. Pilch,
    Identification of the glucagon receptor in rat
    liver membranes by photoaffinity crosslinking,
    Proc. Natl. Acad. Sci. 78:875 (1981).

35. D. Atlas and A. Levitzki, Tentative identification of
    ß-adreno-receptor subunits, Nature 272:370 (1978).

36. A. D. Strosberg, G. Vauquelin, O. Durieu-Trautmann,
    C. Delavier-Klutchko, S. Bottari, and C. André,
    Towards the chemical and functional characteriza-
    tion of the ß-adrenergic receptor, TIBS 5:11 (1980)

37. T. L. Innerarity, E. S. Kempner, D. Y. Hui, and R. W.
    Mahley, Functional unit of the low density
    lipoprotein receptor of fibroblasts : a 100,000-
    dalton structure with multiple binding sites, Proc.
    Natl. Acad. Sci. 78:4378 (1981).

38. C. J. Steer, E. S. Kempner, and G. Ashwell, Molecular
    size of the hepatic receptor for asialoglycopro-
    teins determined in situ by radiation inactivation,
    J. Biol. Chem. 256:5851 (1981).

39. C. Fewtrell, E. Kempner, G. Poy, and H. Metzger,
    Unexpected findings from target analysis of immu-
    noglobulin E and its receptor, Biochemistry 20:
    6589 (1981).

40. A. Doble and L. L. Iversen, Molecular size of benzo-
    diazepine receptor in rat brain in situ : evidence
    for a functional dimer, Nature 295:522 (1982).

41. S. M. Paul, E. S. Kempner, and P. Skolnick, In situ
    molecular-weight determination of brain and peri-

pheral benzodiazepine binding sites, <u>Eur. J.
Pharmacol.</u> 76:465 (1982).

42. L. R. Chang, E. A. Barnard, M. M. S. Lo, and J. O.
Dolly, Molecular sizes of benzodiazepine receptors
and the interacting GABA receptors in the membrane
are identical, <u>FEBS letts</u> 126:309 (1981).

43. M. M. S. Lo, E. A. Barnard, and J. O. Dolly, Size of
acetylcholine receptors in the membrane, an impro-
ved version of the radiation inactivation method,
<u>Biochemistry</u> 21:2210 (1982).

44. S. Uchida, K. Matsumoto, K. Takeyasu, H. Higuchi, and
H. Yoshida, Molecular mechanism of the effects of
guanine nucleotide and sulfhydryl reagent on muscarinic receptors in smooth muscles studied by radiation inactivation, <u>Life Sci.</u> 31:201 (1982).

CONTRIBUTORS

BENFENATI Fabio
    Institute of Human Physiology, University of
    Modena, Via Campi 287, I-41100 Modena, Italy

BRAESTRUP Claus
    A/S Ferrosan, Sydmarken 5, DK 2860 Soborg,
    Denmark

CARBONI Ezio
    Dept. of Pharmacology and Pharmacognosy, Univer-
    sity of Cagliari, Via Diaz 182, I-09100 CAGLIARI,
    Italy

CARRUBA Michele O.
    Dept. of Pharmacology, School of Medicine,
    University of Milan, I-20129 Milan, Italy

COSTA Erminio
    Laboratory of Preclinical Pharmacology, NIMH,
    Saint Elizabeths Hospital, Washington, D.C.
    20032, U.S.A.

ENNA Sam J.
    Dept. of Pharmacology and of Neurobiology and of
    Anatomy, University of Texas Medical School at
    Houston, P.O. Box 20708, Houston, Texas 77025,
    U.S.A.

GAMBALE Franco
    Istituto di Cibernetica e Biofisica, Corso
    Mazzini 20, I-16032 Camogli GE, Italy

GIESING Michael
Labor für Nervengenwebekultur, Institut für Physiologische Chemie der Universität, Nussallee 11, D-5300 Bonn, F.R.G.

GOVONI Stefano
Dept. of Pharmacology and Pharmacognosy, University of Milan, I-20129 Milan, Italy

GUARDABASSO Vincenzo
Institute for Pharmacological Research "Mario Negri", Via Eritrea 62, I-20157 Milan, Italy

HONORE' Tage
A/S Ferrosan, Sydmarken 5, DK-2860 Soeborg, Denmark

LEVITZKI Alexander
Dept. of Biological Chemistry, The Hebrew University of Jerusalem, 91904 Jerusalem, Israel

MARCHETTI Carla
Istituto di Cibernetica e Biofisica, C.so Mazzini 20, I-16032 Camogli GE, Italy

MARCONCINI PEPEU Ileana
Dept. of Pharmacology, University of Florence, Viale Morgagni 65, I-50134 Florence, Italy

MEMO Maurizio
Dept. of Pharmacology and Pharmacognosy, University of Cagliari, Via Diaz 182, I-09100 Cagliari, Italy

MUNSON Peter J.
Laboratory of Theroretical and Physical Biology, National Institute of Child Health and Human Development, NIH, Bethesda, Maryland 20205, U.S.A.

NIELSEN Mogens
Sct. Hans Mental Hospital, 4000 Roskilde, Denmark

PEPEU Giancarlo
    Dept. of Pharmacology, University of Florence,
    Viale Morgagni 65, I-50134 Florence, Italy

ROBELLO Mauro
    Istituto   di   Scienze   Fisiche,   Università   di
    Genova, Viale Benedetto XV/5,   I-16132   Genova,
    Italy

SPANO Pierfranco
    Dept. of Pharmacology and Pharmacognosy, Univer-
    sity of Cagliari, Viale Diaz 182,   I-09100
    Cagliari, Italy

SCHLEGEL Werner
    Fondation pour   Recherches Medicales, Univ. of
    Geneva, 64, Avenue de la Roseraie, CH 1211 Geneve
    4, Switzerland

TRABUCCHI Marco
    Dept. of Pharmacology, University of Brescia,
    I-25100 Brescia, Italy

USAI Cesare
    Istituto di Cibernetica e Biofisica, C.so Mazzini
    20, I-16132 Camogli GE, Italy

UZUMAKI Hiroya
    Dept. of Pharmacology and Pharmacognosy, Univer-
    sity of Milan, I-20129 Milan, Italy.